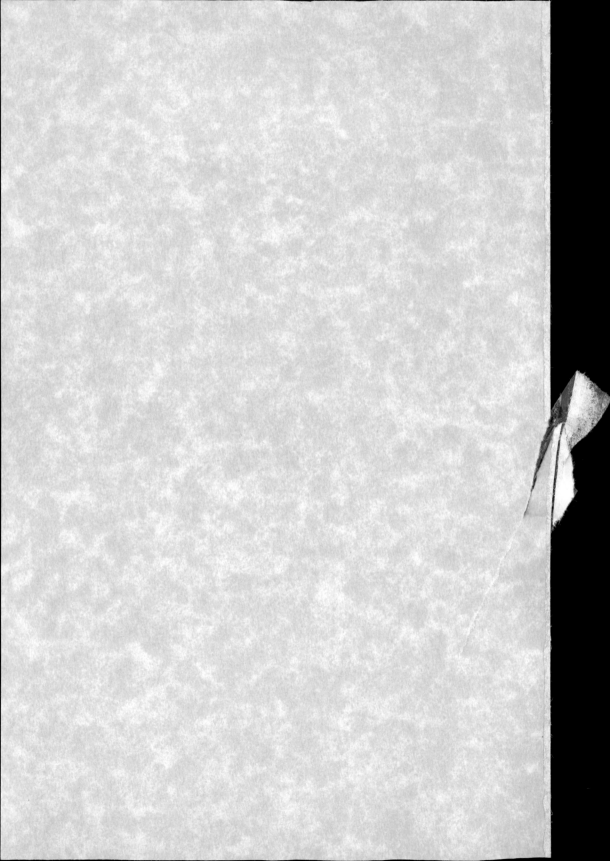

CONTENTS

Foreword   v

Introduction   The Dissolution of a Fundamentalist   1

CHAPTER 1   The Lie Among Us   19

CHAPTER 2   A Tale of Two Books   43

CHAPTER 3   Darwin's Dark Companions   63

CHAPTER 4   The Never Ending Closing Argument   85

CHAPTER 5   The Emperor's New Science   121

CHAPTER 6   Creationism Evolves into Intelligent Design   145

CHAPTER 7   How to be Stupid, Wicked, and Insane   165

CHAPTER 8   Evolution and Physics Envy   185

CONCLUSION   Pilgrim's Progress   207

Acknowledgments   222

Notes   224

Index   240

SAVING DARWIN

How to Be a Christian and Believe in Evolution

Karl W. Giberson

HarperOne
An Imprint of HarperCollins*Publishers*

HarperOne

HarperCollins books may be purchased for educational, business, or sales promotional use. For information please write: Special Markets Department, HarperCollins Publishers, 10 East 53rd Street, New York, NY 10022.

HarperCollins Web site: http://www.harpercollins.com

HarperCollins®, ♣®, and HarperOne™ are trademarks of HarperCollins Publishers.

A Giniger Book
The K.S. Giniger Company
1045 Park Avenue
New York, NY 10028

FIRST EDITION

Library of Congress Cataloging-in-Publication Data

Giberson, Karl.
Saving Darwin : how to be a Christian and believe in evolution / Karl W. Giberson. —1st ed.
p. cm.
Includes bibliographical references (p.) and index.
ISBN: 978-0-06-122878-0 (hardcover)
1. Evolution (Biology)—Religious aspects—Christianity.
2. Creationism. 3. Darwin, Charles, 1809–1882. I. Title.
BT712.G53 2008
231.7'652—dc22
2007034605
08 09 10 11 12 RRD (H) 10 9 8 7 6 5 4 3 2 1

FOREWORD

Americans live, according to the lyrics of their national anthem, in the land of the free and the home of the brave. They also live in a land that hosts one of the great paradoxes of our time. Many of its citizens have faith in science and technology to solve society's problems, but many others have faith in a literal interpretation of the book of Genesis that is utterly in conflict with what science tells us about our own origins.

The science-religion conversation is often not a friendly debate. A spate of angry new books denouncing religious faith has appeared, some of them penned by atheist biologists who use evolution as a club to berate believers. On the other side of the great divide, the Intelligent Design (ID) movement presses on with its challenge to evolution's ability to explain "irreducibly complex" structures in living organisms, despite lack of any meaningful support in the scientific community and a recent stunning court defeat of the plan to teach ID as an alternative to evolution in the school system. As perhaps the strangest development of all, a "creation museum" has opened just outside Cincinnati, depicting humans frolicking with dinosaurs, despite overwhelming scientific evidence that they were separated in history by more than sixty million years. What's going on here? How can the most advanced technological country in the world also be home to such antiscientific thinking?

Some have dismissed this as an inevitable consequence of the fact that Americans take their religion seriously. In that context, they say that this is just one more chapter in a perpetual and irreconcilable battle between science and faith, arguing that these worldviews are simply incompatible and that individuals have to make a choice about which to believe in. But, as Karl Giberson ably describes in this much-needed book, that would be a misrepresentation of the facts. In reality, science and religion have generally coexisted quite comfortably until about a century or so ago. Copernicus, Kepler, and Galileo were all firm believers, and Newton wrote more words about biblical interpretation than he did on mathematics and physics. Clearly the greatest threat to that harmony has been the arrival of Darwin's theory of evolution, but even that development was not initially seen by leaders of the Christian church as all that threatening to their worldview after publication of *On the Origin of Species.*

Giberson has provided a critical service by leading us carefully through a series of historical events that began in the late nineteenth century and led to the current culture wars. These events stretch from Ellen White's Seventh-day Adventist visions of creation, to the birth of fundamentalism as a response to a liberal form of Christian theology that actually denied the divinity of Christ, to the human misery wrought by those who misused Darwin's theory to justify oppressive social changes, to the ill-conceived but still widely embraced *The Genesis Flood* of Henry Morris, which proposed a scientific basis for a very young earth.

Giberson's carefully documented history provides a sobering response to the claims of those who think that the current controversy can be quickly resolved. Just as with other great world conflicts, such as the current war in the Middle East, we will be forever doomed to disappointment in an effort to find peace and harmony if we don't understand how we got to this contentious juncture.

C. S. Lewis, the great proponent of a rational approach to Christian faith, led the Socratic Club at Oxford more than half a century ago, and the motto of the group was "to follow the argument wherever it leads." *Saving Darwin* is in that distinguished tradition. We should all be able to agree, believers and nonbelievers alike, that finding the truth is our task. We may disagree about how to interpret some of the facts, of course, but we cannot dismiss them as just inconvenient.

Here are some true statements that cannot be ignored:

Darwin's theory of evolution has been overwhelmingly supported by evidence from a wide variety of sources. Those include the increasingly detailed fossil record, but even more compelling evidence now comes from the study of genomes from many organisms, providing much more proof of common descent (including *Homo sapiens*) than Darwin could have dreamed of. Given such oddities in our own DNA record as pseudogenes and ancestral chromosome fusions, special creation of humans simply cannot be embraced by those familiar with the data, unless they wish to postulate a God who intentionally placed misleading clues in our own DNA to test our faith.

Alternatives to evolution such as young- or old-earth creationism and intelligent design find almost no support in the scientific community. Although many nonscientist Christians have been taught to embrace one or another of these alternatives as a means of opposing the perception that evolution is godless, the God of all truth is not well served by lies, no matter how noble the intentions of those who spread them.

On the other hand, a purely naturalistic worldview can be justly criticized as narrow and impoverished. Science must forever remain silent on questions such as: "What is the meaning of life?" "Is there a God?" "Do right and wrong have any real meaning?" and "What happens after we die?" And yet surely those are profoundly important questions that we humans should be trying to answer. Only a spiritual worldview can help us here.

The good news is that there is a harmonious solution at hand. Many working scientists, including Giberson and myself, find no conflict in both embracing the conclusion that evolution is true and seeing this as the means by which God implemented his majestic creation. In that synthesis of the natural and spiritual perspectives we have found much joy and peace, where our increasingly detailed understanding of the molecules of life only adds to our awe of the Creator. Put in that framework, DNA is essentially the language God used to speak us and all other living things into being.

Yet the culture wars continue. And if some resolution is not found soon, we will all be the losers. Would that we could return to the exhortations of theologians like Benjamin Warfield, who wrote these words in the late nineteenth century, fully aware of the significance of Darwin's theory and unafraid of its consequences for the future of the Christian faith:

> We must not, then, as Christians, assume an attitude of antagonism toward the truths of reason, or the truths of philosophy, or the truths of science, or the truths of history, or the truths of criticism. As children of the light, we must be careful to keep ourselves open to every ray of light. Let us, then, cultivate an attitude of courage as over against the investigations of the day. None should be more zealous in them than we. None should be more quick

to discern truth in every field, more hospitable to receive it, more loyal to follow it, whithersoever it leads. (From B. B. Warfield, *Selected Shorter Writings* [Phillipsburg, NJ: PRR Publishing, 1970, pp. 463–65.])

Saving Darwin is a powerful contribution to this critically important effort to seek an enlightened and worshipful peace. With clearly presented statements of truth like those within these pages, together with a shared confidence that scientific discoveries about nature can hardly threaten nature's Creator, perhaps we have a chance in this century to develop a new Christian theology that celebrates God's awesome creation, unafraid of what science can tell us about the details. Then perhaps we can get beyond these destructive battles to focus on the real meaning of Christianity. That actually has little to do with alternative creation stories and everything to do with God's love as demonstrated most profoundly in the life, death, and resurrection of Jesus Christ.

Francis S. Collins, M.D., Ph.D.

THE DISSOLUTION OF A FUNDAMENTALIST

I n 1975 I left my home in maritime Canada to attend Eastern Nazarene College on Boston's historic south shore. Among my prized possessions, as I nervously traded the potato fields for the big city, were dog-eared copies of Henry Morris's classic texts of scientific creationism and Christian apologetics, *The Genesis Flood* and *Many Infallible Proofs*.[1]

Morris, who passed away in early 2006 as I was writing these words, was one of my boyhood heroes. As Willie Mays had inspired me to play center field, and Gordon Lightfoot the guitar, so Morris inspired me to master the art of Christian apologetics, to be, in the immortal words of St. Paul, "not ashamed of the testimony of our Lord." Morris, a giant of American fundamentalism, profoundly influenced religion in twentieth-century America, an influence that extended undiminished into much of Canada as well.

My childhood experiences in center field convinced me that, although I had mastered Mays's famous basket catch, baseball held no future for me. The great gulf between Gordon Lightfoot's guitar playing and my own confirmed that I would never make a living in folk music. But I was good at math and science—and arguing—and it looked as though I might follow in Morris's footsteps and become a Christian apologist. I was particularly enamored with Morris's eloquent and scientifically informed defense of the Genesis creation story and his clear-headed refutation of Darwinian evolution. I planned to major in physics, get a Ph.D., and go to work at Morris's recently created Institute for Creation Research in San Diego, where I would join those noble fundamentalist warriors as they stormed the ramparts of evolution and rescued the Genesis story of creation.

Like many young people raised in fundamentalist churches, I had been captured by the promise of scientific creationism, which Morris had launched in the early 1960s with the publication of his remarkable book *The Genesis Flood*. In that classic and impressively technical work, Morris and his coauthor, Old Testament scholar John C. Whitcomb, argue persuasively that the Bible and the Book of Nature agree that the earth was created in its present state about ten thousand years ago. The 518-page volume, which has sold over a quarter million copies and is still available in its forty-fourth printing, had enough footnotes, graphs, and pictures to convince any intellectually oriented fundamentalist that there was no reason to take evolution seriously. Readers could rest assured in the knowledge that Darwin's theory was deeply flawed, without empirical support, and on the verge of collapse. A few celebrated and highly publicized defections from the evolutionary camp illustrated the magnitude of the problem and suggested that this was an opportune time to join the war against Darwin's evil theory. In stark contrast to the failing fortunes of evolution, Whitcomb and Morris argued persuasively that the biblical creation story became increasingly credible as scientific evidence accumulated.

My first year at Eastern Nazarene College, which wasn't the fundamentalist haven I had anticipated, was troubling. Away in a strange new city, homesick for the rolling hills of the beautiful St. John River Valley I had left behind in New Brunswick, and without close friends, I struggled in the classroom. My Bible professor assaulted my literalist reading of Genesis, suggesting that Genesis should be read as poetry rather than science, a liberal heresy that Morris had warned me I might encounter. To make matters worse, the science faculty—despite claiming to be Christians—all seemed to accept evolution. Even my fellow students, at least in the science division, had limited interest in the creationist cause to which I had heroically dedicated myself.

These experiences steeled my resolve to stay the course. My extensive reading in fundamentalist apologetics and scientific creationism—and my enthusiasm for arguing—gave me confidence I was right. I could quote credentialed biblical scholars who understood that Genesis was more than poetry and that Christian theology would come apart if Genesis was not read literally. I had books by real scientists refuting evolution with solid arguments that, strangely, many of my professors did not know. The literature buttressing my position was extensive, my authorities were unassailable, and someday I too would have the credentials to speak with authority on this topic.

During my freshman year I attended a creationist event at Boston University, where Duane Gish, the premier and highly polished creationist debater, humiliated his inarticulate and unprepared opponent, who utterly failed to defend evolution. A vision of myself in that same role, perhaps a decade hence, further inspired me. At the end of the year I had the good fortune to meet the grand old man of creationism himself—Henry Morris—at a local church, where he was giving a Saturday seminar on creation. I chatted with him afterwards, and he encouraged me on my course, suggesting that I follow through on my plans to earn a Ph.D. in physics and then contact him at the Institute for Creation Research for a possible research position. He signed my well-worn copy of his manifesto, *Many Infallible Proofs,* inscribing the following biblical reference, 2 Timothy 1:7–9:

> For God hath not given us the spirit of fear; but of power, and of love, and of a sound mind. Be not thou therefore ashamed of the testimony of our Lord, nor of me his prisoner: but be thou partaker of the afflictions of the gospel according to the power of God; who hath saved us, and called *us* with an holy calling, not according to our works, but according to his own purpose and grace, which was given us in Christ Jesus before the world began.[2]

I WAS A TEENAGE FUNDAMENTALIST

Scientific creationism, the idea that the biblical story of creation rests on solid scientific evidence, is an integral part of the fundamentalist worldview that inspired me as a teenager. This understanding of Christianity starts with the assumption that the Bible is completely without error of any kind, having essentially been written by God. Scientific statements in the Bible are completely accurate, and historical references are utterly reliable. All statements on all topics are absolutely trustworthy in all respects. This is the fundamentalist creed, learned at mother's knee, reinforced in Sunday school, to be defended at all costs.

God inspired the biblical authors in such a way that their writings would be indistinguishable from dictation directly from God. God is thus the *author* of the Bible, and the "writers" are little more than scribes. Fundamentalist preachers quote Scripture constantly, rarely introducing it with anything other than "the Bible says" or "God says." This view of Scripture gives the Bible both an extraordinary authority and a complete unity of perspective. It has one author and no errors. Complex arguments can thus be securely developed by lifting bits of text from widely disparate books of the Bible and combining them, just as geometrical proofs can be constructed by combining axioms and theorems. If God wrote the entire Bible, then it is one long coherent message.

God provided the Genesis creation story so that we might understand our origins. In this account we read that God created a perfect world, with no sin, no death, and great harmony between his creatures and himself. Under the temptation of Satan, the first human couple, Adam and Eve, sinned—of their own free will—bringing death, suffering, and destruction into the world. If they had not sinned, they would still be alive, listening to music on their iPods and enjoying millions of great-grandchildren. This is the clear meaning of the text, taken at face value. Any other reading implies that God created an imperfect world and that the evils of death and suffering were part of his original creation.

Such dramatic and deeply counterintuitive elements are common in the fundamentalist reading of the Genesis creation story. The first appearance of sin in a perfect creation was a catastrophic transformation, like a crack in a magnificent glass window or a beautiful vase. Sin completely changed the *physical* as well as the *moral* structure of the world, introducing a major "break" in natural history. Women's bodies were altered so childbirth would be painful. The ecology changed so growing crops would be hard work. Plants developed thorns, and helpful bacteria turned into sinister parasites, inflicting disease on their hosts. Elsewhere in the Bible we read that all of creation "groans" under a universal curse that an enraged God placed on the creation because of the sin of Adam and Eve.[3] Many scientific creationists identify this curse with the physicists' famous second law of thermodynamics, that mysterious statement that nature constantly grows ever more disordered as time passes. What better explanation for the origin of this law than the sin of Adam and Eve?

THE END OF CREATION

At the end of the creation story in Genesis, God rests. Whatever processes were used to "create" shut down on the sixth day of creation and are no longer a part of the natural order. Science thus has no access to these processes and is limited to studying the stable, status-quo, postcreation patterns of nature. It follows that there can really be no "science" of origins, and we should not expect to understand the various mechanisms—all of them supernatural—that God used to create the world. Secular scientists err in attempting to understand origins by inspection of the fossil record and geological history. The record that the geologists and paleontologists are reading to recreate the natural history of our planet is not the story of our origins; it is, in fact, nothing more than the residue of Noah's great flood.

The flood story is a central underpinning of scientific creationism. Genesis says that the human race, about four thousand years ago, had become so wicked it had to be annihilated. God wiped out almost all humanity with a flood—a global cataclysm that completely reshaped the surface of the earth. This flood laid down virtually all the fossil strata we find today and completely contoured the surface features of the earth, from the Grand Canyon to Mt. Everest. Tectonic activity thrust up mountains. Receding floodwaters carved out canyons, both grand and small. The flood scoured off any prior earth history, like a bulldozer removing an ancient forest to make room for a parking lot.

The classic text by Whitcomb and Morris, *The Genesis Flood*, marshals scientific evidence for this biblical story, arguing that it provides a better explanation for the fossil record and the surface geology of the earth than the conventional scientific account arising from the erroneous assumption by misguided scientists that the earth is billions of years old. *The Genesis Flood* also argues effectively that the Bible intends us to take the flood story literally and understand it as a global, rather than local, event. After the floodwaters receded, God promised Noah that he would never again flood the earth. He placed a rainbow, for the first time, in the heavens as a sign of his promise. The laws of physics changed at this time—about four thousand years ago—to enable rainbows.

Whitcomb and Morris argue convincingly that the scientific and biblical witnesses to these historical accounts agree perfectly. So why, I wondered, does such widespread opposition exist within the scientific community? How can it be that the entire academic community of geologists rejects the worldwide flood of Noah and claims the earth is billions of years old? Why are biologists so blind to the simple truth that God created the world in six days? Why do physicists and astronomers propose so many ideas— from radioactive dating to stellar evolution to the big bang—that suggest the universe is ancient? Why do so many biblical scholars—who claim to be

Christians—reject the biblical witness to all of this? Why do theologians say that none of this matters?

Morris's answers to these questions are simple. Human beings, he explains, are fallen, sinful creatures, easily deceived by Satan. Blind to God's truth, secular scientists and liberal scholars of religion are unknowingly doing the will of the devil. The existence of such a widespread conspiracy to destroy the simple truths of Genesis demands nothing less than just such a comprehensive explanation. Satan has deceived the scientific community, and a great many Christians as well.

Apparently, I wasn't the only reader convinced by the arguments of Whitcomb and Morris. A 2004 CBS poll revealed that over half the population of the United States accepts the biblical creation story, many of them embracing the exact version Whitcomb and Morris presented a half century ago.[4] This position is thoroughly at odds with almost *all* the relevant scholarship of the past century. Today I would describe this view as sophomoric in the most literal sense of the word, which it certainly was for me, as I watched it wilt over the course of my sophomore year in college. By the middle of that critical year I was sliding uncontrollably down the slippery slope that has characterized religion since it began the liberalizing process just over a century ago.

THE EVOLUTION OF A FUNDAMENTALIST

An interesting concept in evolutionary theory is the pompous-sounding *ontogeny recapitulates phylogeny*. Originally proposed by the German evolutionist Ernst Haeckel in 1866, this idea claims that the development of the embryo of a species—its *ontogeny*—is a fast-forward version of its entire evolutionary history—its *phylogeny*. The sequence of developmental steps through which an embryo passes as it matures—in mother's womb, for humans—is a mirror of the developmental steps through which the species has passed in the course of its evolution over millions of years.

Scientists today reject much of Haeckel's once influential idea. Nevertheless, the concept provides a marvelous description of the process I went through in my sophomore year of college as I evolved rapidly from the simple intellectual life-form called *Homo fundamentalis* to something more complex, in the process passing rapidly through the various intermediate forms that emerged in the decades since Darwin.

As I studied science and mathematics, I began to doubt that science could have gotten everything as thoroughly wrong as the creationists suggested. The simple physics of radioactivity, widely used to date rocks, provides a characteristic example. Many different ways exist to date the earth, and almost all of them agree that the earth is billions, not thousands, of years old. If the earth was really just a few thousand years old as the Bible seemed to indicate,[5] why would God plant evidence to trick us into thinking it was billions of years old?

Just as my counterparts in the eighteenth and nineteenth centuries struggled to reconcile the new geology of their day with the Bible, I tried at first to play with different, but still literal, readings of Genesis. Maybe I could salvage the Genesis story by reading the "days" of creation as long periods of time. But this didn't seem reasonable. The Bible says, "In the beginning God created the heaven and the earth," while science says the earth appeared some nine billion years *after* the universe began. Furthermore, God created the sun on the fourth day, *after* the vegetation, which presumably needed the sun to survive. If the third day was a billion years long, the vegetation would have been long gone before the photosynthesis of the fourth day ever got started.

Each new question made things more complicated. A billion-year-old earth demands that we reinterpret "the fall." As long as Adam and Eve appeared in the same week as everything else, it was at least possible that their "sin" brought unintended death and suffering into the world. But now it appears that death and suffering had been present for a billion years with entire species going extinct long before humans appeared. Why would God create species only to have them go extinct long before Adam even had time to name them? Was this the same God who would later *preserve* every species on the planet by having Noah build an ark to rescue them from the flood? If extinction was normal, why did we need an ark? What, exactly, were the implications of the fall?

The acceptance of an ancient earth brings other troubles. If we take the geological record seriously, we confront fossils of what look like humans in rock strata more than a hundred thousand years old. And these fossils look as if they belong to a species that evolved from similar, earlier species. If we line up all these species in historical order, we have what certainly looks like a compelling narrative of human evolution from subhuman ancestors. Where in this history do we place Adam and Eve? No logical place appears in the unbroken sequence of human evolution for the famous residents of the Garden of Eden. And where, exactly, *was* the Garden of Eden? The Genesis story says that God placed an angel at the entrance to keep people out, which certainly implies that it was to continue even after Adam and Eve were expelled. We have no record of God closing it down. If God didn't destroy Eden, where is it now?

Doubts about the historicity of Adam and Eve and the Garden of Eden make it hard to read the creation stories without asking additional difficult questions. And fundamentalists in the midst of their theological breakdowns look in vain to contemporary biblical scholarship for help. Al Truesdale, my freshman Bible professor, had offered many helpful suggestions just a year earlier, bless his heart, but I had rejected all of them. They now came rushing back to haunt me. I found myself in an uncomfortable alternate reality that was a strange and darkened mirror image of the fundamentalist world I had inhabited for my entire life.

Fundamentalists find a satisfying harmony between science, as they understand it, and the Bible, as they interpret it. Their "science" is scientific creationism, which gathers evidence for the Genesis creation story. Their approach to the Bible is biblical literalism, which reads the text in the simplest way possible. These approaches reinforce each other and make the whole greater than the sum of the parts. But real science, which I was studying in college, and contemporary biblical scholarship, which religion majors were studying, conspire in such a way that the whole becomes *less* than the parts. The Genesis story of creation loses all contact with natural history and starts to look strangely like an old-fashioned fairy tale that might teach a lesson, but certainly makes no claim to historicity.

I learned, for example, that the word we translate as "Adam" in our English Bibles simply means "man" in Hebrew. And "Eve" means "woman." I began to wonder how an old story about a guy named "Man" in a magical garden who had a mate named "Woman" made from one of his ribs could ever be mistaken for actual history. And yet this was exactly what I had believed just one year earlier. Talking snakes, visits from God in the evening, naming the animals—the story takes on such a different character the moment one applies even the most basic literary analysis. The literalist interpretation I had formerly embraced and defended so vigorously began to look ridiculous, as did the person I had been just one year earlier.

THE JENGA TOWER

I would have liked to find some simple alternative reading of Genesis to re-place the literalist interpretation, but, if one existed, I certainly couldn't find it. I turned with some optimism to religion scholars, but found they had lit-tle to offer. Some of them strangely insisted on the historicity of *some* por-tions of the Genesis story, while allowing that much of it was not historical. The fall, for example, was sometimes an important part of elaborate theo-logical systems, serving the critical function of getting God off the hook for a creation filled with so much suffering. So even though Adam and Eve were not actual characters themselves and Eden was not a real place, they at least represented *something* historical. Once upon a time human beings did *some-thing* to ruin God's perfect creation, and this is where it all went wrong.

I was now wearing scientific spectacles almost all the time, and these ex-planations looked a little too convenient to me. Some theologians, for exam-ple, liked the way that Paul's reference to Jesus as the "second Adam" drew a provocative connection between the fall and redemption (1 Cor. 15:45). The first Adam made the mess; the second Adam cleaned it up. I could never see, though, how theologians could be so comfortable with a mythical interpretation of Eden, but insist on an important historical role for its first resident. Paul's "first Adam" was indeed the original sinner, but he didn't live in the Garden of Eden, he didn't name all the animals, and he may or may not have been married to Eve.

Further complicating my struggles, the religion scholars I consulted were quite accepting of evolution. An Old Testament scholar with a Ph.D. from Boston University assured me that "Genesis was never intended to be read literally." He and his colleagues had made their peace with evolution, appar-ently as toddlers, and had been at peace about this ever since. They were surprisingly disinterested in the struggles of those who, like me, were try-ing to hold on to some version of their childhood faith, while portions of its foundations were slowly removed, like the pieces of a Jenga tower that may or may not come crashing down as once extracts the tiny logs.

THE UNIVERSAL ACID OF DARWINISM

Tufts University philosopher Daniel Dennett describes evolution as a "uni-versal acid." With undisguised glee he outlines how evolution, which he calls

"Darwin's dangerous idea," eats through and dissolves the foundations of religion. The theory of evolution, which he thinks is the greatest idea anyone ever had, destroys the belief that God created everything, including humans. "Darwin's idea," he writes with approval, "eats through just about every traditional concept, and leaves in its wake a revolutionized worldview."[6]

Acid is an appropriate metaphor for the erosion of my fundamentalism, as I slowly lost my confidence in the Genesis story of creation and the scientific creationism that placed this ancient story within the framework of modern science. Dennett's universal acid dissolved Adam and Eve; it ate through the Garden of Eden; it destroyed the historicity of the events of creation week. It etched holes in those parts of Christianity connected to these stories—the fall, "Christ as second Adam," the origins of sin, and nearly everything else that I counted sacred. I discovered, however, that this was about where Dennett's acid ran out of steam (or whatever acid runs out of when it stops dissolving everything). The acid of evolution is not universal, and claims that evolution "revolutionizes" our worldview and dissolves every traditional concept are exaggerated.

For starters, what exactly does evolution have to do with belief in God as creator? It rules out *certain* mechanisms that God might have used to create the world, but others remain. God apparently did not create the entire universe and everything in it over the course of a few busy days ten thousand years ago. Neither Rome nor the universe was built in a day. But saying that Rome was not built in a day does not imply that Rome was not built or that Rome did not have builders. The acid of evolution dissolves the claim that God created the world a few thousand years ago, but does nothing to the claim that God may have taken billions of years to create or that God even continues to work as creator.

Creation, I hasten to point out, is a *secondary* doctrine for Christians. The central idea in Christianity concerns Jesus Christ and the claim that he was the Son of God, truly divine and truly human. This extraordinary idea implies the strange notion that the creator of the entire universe chose to enter the human race in the person of an itinerant preacher from Galilee. From its beginnings Christianity had to defend itself against charges that this was a ridiculous idea. Some of the most influential early church fathers were quite clear that the claims of Christianity were, indeed, absurd, but this did not mean they were not true. A second-century theologian named Tertullian said he believed in the divinity of Jesus partly because it *was* absurd.[7]

Most thoughtful Christians, myself included, wonder about exactly how it could be that God entered the human race in the person of Jesus—the historical event called the Incarnation. Over the centuries many have been simply unable to believe that this claim was even sensible. Today thinking Christians everywhere struggle with this belief and what it means. Many have asked God for more faith, to keep doubt at bay or reestablish a foundation for belief. Darwin's theory of evolution adds *nothing* to the complexities and challenges of believing in the Incarnation. It didn't take Darwin to make Christianity offensive, complex, and intellectually challenging. The arguments against the incarnation have been around for two thousand years, which is why Christianity is described as a *faith,* not as the conclusion of a logical argument.

Christianity merges the Incarnation with the belief that Jesus rose from the dead. Christ's Resurrection offers hope that we too can have eternal life and one day be united with God. Human skepticism regarding these claims is hardly new. The contemporaries of Jesus found this hard to believe, and many of them, including the infamous "doubting Thomas," had to be convinced by more than hearsay. Human beings, including Jesus, may have evolved over billions of years, or they may have been created a few thousand years ago. The Resurrection is equally implausible in either case. Dennett's universal acid of evolution does nothing to eat away at this central Christian belief. The "acid" of logic and reason was hard at work on this before the New Testament was even penned.

Christianity, as its name suggests, is *primarily* about Christ. To be sure, different ideas about Christ exist across the spectrum of Christian belief. But these beliefs, rather than creationist assertions, are the heart and soul of Christianity. And these beliefs are not threatened by Darwin's dangerous idea. Evolution does, however, pose two challenges to *secondary* Christian beliefs: the *fall* of humankind, and the *uniqueness* of humankind.

DISSOLVING THE FALL

Clearly, the historicity of Adam and Eve and their fall from grace are hard to reconcile with natural history. The geological and fossil records make this case compellingly. Nevertheless, scholars have proposed many convoluted and implausible ways to resolve these tensions in the past couple centuries. One could believe, for example, that at some point in evolutionary history

God "chose" two people from a group of evolving "humans," gave them his image, and then put them in Eden, which they promptly corrupted by sinning. But this solution is unsatisfactory, artificial, and certainly not what the writer of Genesis intended. Nor does any historical evidence suggest this interpretation. This modification also does absolutely nothing to support the idea that death did not exist in the world before sin. We must concede that the acid of evolution has indeed eaten away the literal part of this story, but I would argue that the most important part of the story remains untouched.

The idea at the center of the fall is human sinfulness. Human beings are sinful creatures, and many of us are really quite dreadful. Even the best of us dare not lay claim to anything even approaching perfection. G. K. Chesterton once quipped that the sinful nature of humans was the only Christian doctrine that we could confirm empirically.[8] The classic story of the fall is best understood as a powerful statement that we are, when all is said and done, sinful creatures.

But what, exactly, does it mean to be *sinful*? Various theological interpretations exist, some more compelling than others. But when the rubber hits the road, *sinfulness* is mainly *selfishness*. We put ourselves ahead of others and ahead of God. We advance our own agenda as if that is all that matters.

Evolution says some interesting things about selfishness. Selfishness, in fact, drives the evolutionary process. Unselfish creatures died, and their unselfish genes perished with them. Selfish creatures, who attended to their own needs for food, power, and sex, flourished and passed on these genes to their offspring. After many generations selfishness was so fully programmed in our genomes that it was a significant part of what we now call human nature.

But an interesting tension exists in human nature. As incurably selfish as we appear to be, we also possess an innate altruism. Human beings are easily capable of actions that benefit others at their own expense—from taking a pie to a new neighbor, to giving money to charities, to risking one's life to save a child. Although altruism is scientifically harder to understand than selfishness, it remains clear that humans are a powerful mix of selfish and unselfish tendencies.

So where does sin originate? In the traditional picture, sin originates in a free act of the first humans: God gave humans free will and they used it to contaminate the entire creation. That was the risk God took in creation. But now we have a new and better way to understand the origins of sin. We start by enlarging our own troublesome "freedom" to include nature. In the same way that we possess a genuine freedom to explore possibilities, nature has freedom as well, although not a conscious freedom, of course. Physicists enshrine this insight in the Heisenberg Uncertainty Principle, which accords a degree of genuine "freedom" to particles like the electron.

If nature, in all its many processes, is "free" to explore pathways of possibility, then the evolutionary process would predictably lead to creatures with pathological levels of selfishness. Creatures inattentive to their own needs would not have made it. By these lights, God did not "build" sin into the natural order. Rather, God endowed the natural order with the freedom to "become," and the result was an interesting, morally complex, spiritually rich, but ultimately selfish species we call *Homo sapiens*. This is an entirely reasonable theological speculation, at least by my amateur standards. It brings the Christian doctrine of the fall into the larger picture of an extended creation. Humankind did not appear all at once, and neither did sin.

DISSOLVING THE UNIQUENESS OF HUMANKIND

Once we accept the full evolutionary picture of human origins, we face the problem of human uniqueness. The picture of natural history disclosed by modern science reveals human beings evolving slowly and imperceptibly from earlier, simpler creatures. None of our attributes—intelligence, upright posture, moral sense, opposable thumbs, language capacity—emerged suddenly. Every one of our remarkable capacities must have appeared gradually and been present in some partial, anticipatory way in our primate ancestors. This provocatively suggests that animals, especially the higher primates, ought to possess an identifiable moral sense that is only *quantitatively* different from that of humans. Not surprisingly, current research supports this notion.

Scientists who have spent enough time with primates, especially in natural settings, are continually struck by their sophistication. In his remarkable books on primates, Emory University primatologist Frans de Waal describes primate behaviors that, were they associated with humans, would suggest a well-defined sense of right and wrong, cruelty and kindness, loyalty and manipulation. A remarkable bonobo named Kuni, to recount one example, saw a starling hit a glass wall and plummet to the ground. Kuni carefully picked up the stunned bird, set it on its feet, and waited with apparent concern for it to fly. When it didn't fly off on its own, Kuni picked up the bird and carried it carefully to the top of a large tree. Wrapping her legs around the tree to free both hands, Kuni spread the wings of the bird and released it, only to watch it flutter to the ground. Kuni then stood watch over the bird for a good portion of the day until it finally recovered and flew off on its own.[9] This story is close enough to that of the good Samaritan to make it hard to treat morality as a purely human attribute. And we have records of countless other examples of similar animal behaviors.

Primates have learned enough language to communicate with over a hundred symbols. They can do simple math, punching a key for "3" when they see three candies in a bowl. Primate "societies" are home to such typically human behaviors as male competition, the bullying of nerds, and female solidarity. Researchers find traits like loyalty, jealousy, and generosity among primates and other species as well. Anthropologists have even observed what look like collective spiritual gatherings of primates, in which a group of chimpanzees will gather to watch, in silence, a beautiful sunset, dispersing after the event when a leader signals it is time to go. The large number of human traits that appear in primate societies is intriguing and sobering, especially as we contemplate the ongoing threat that our activities pose to them.

Does the "acid" of our evolutionary kinship with the primates dissolve anything of importance to Christian theology? I am not convinced that it does.

The tricky issue for Christianity is teasing out which biblical and theological claims derive from a mistaken picture of science and which are central to the ongoing vitality of the faith. Until recently just about everyone in all cultures perceived a great *qualitative* distinction between humans and the higher primates. Certainly the biblical writers and the formative thinkers of the Christian tradition could not have anticipated what we have learned from primate studies in the past few decades. So we may suppose that they would frame their religious understanding in exclusively human terms. In the same way Christian cosmology was developed with the earth at the center of the universe, because that was the best understanding at the time.

Speculations such as these are above my pay grade, of course, and best left to theologians. Still, I find no compelling reason to think that the central message of Christianity is incompatible with humanity's kinship with the rest of the animal world. In fact, this continuity with the animal world may place increasing theological significance on the welfare of animals and ecological responsibility.

THE VIEW FROM OUTSIDE

Many informed and careful Christian thinkers have made their peace with evolution and found ways to incorporate its central insights into their theology. Coming from conservative evangelical traditions are physicist Howard Van Till, in the Reformed tradition, formerly of Calvin College, and biologist Darrel Falk, from the Wesleyan tradition, who currently teaches at Point Loma Nazarene University. These respected thinkers ventured into the troubled waters of evolution and wrote popular books in an effort to bring their respective denominations out of the nineteenth century. [10] Both are committed Christians with stellar records of serving at their respective denominational colleges. Yet powerful, but deeply uninformed fundamentalists who wanted them censured assaulted their works.

Recently the head of the Human Genome Project and one of America's most visible scientists, Francis Collins, has endorsed the idea that evolution is compatible with Christianity. Collins, who converted from atheism to evangelical Christianity after reading C. S. Lewis's *Mere Christianity,* wrote *The Language of God: A Scientist Presents Evidence for Belief.*[11] In that influential book Collins stakes out a middle ground for evolution between the dogmatisms of atheistic materialism and fundamentalist creationism.

The Roman Catholic tradition currently has a significant dialog with science, and the Pontifical Academy of Science numbers many leading scientists, including evolutionists, among its members. This dialog has allowed Catholicism to avoid much of the anti-evolutionary frenzy that rained down on Falk and Van Till. Out of this tradition come Brown University biologist Ken Miller and Georgetown University theologian John Haught. Miller's 1999 *Finding Darwin's God* became something of a classic and its author an important public intellectual and symbol of the integration of evolution and Christianity.[12] Haught has written several books in this area, the most important of which is *God After Darwin,* a tweaking of traditional Catholic theology in response to evolution.[13]

In England, two influential theologians, Alister McGrath and Keith Ward, have penned several popular works apiece integrating evolution and Christian theology. McGrath holds the chair of Professor of Historical

Theology at Oxford University and Ward is the Emeritus Regius Professor of Divinity at Oxford, the most prestigious theological posting in the Anglican Church. McGrath has written the three-volume *Scientific Theology*, inaugurating a major project to reformulate Christian theology in light of recent scientific developments, particularly evolution.[14] Ward's *God, Chance and Necessity* offers helpful ways to reconcile evolution with belief in the doctrine of creation.[15]

Philosopher Michael Ruse has also made an interesting contribution. A prolific author, Ruse has been a fixture in America's creation–evolution controversy since he testified for the American Civil Liberties Union (ACLU) at the Arkansas "Scopes II" trial in 1981. In response to claims that the truth of evolution entails the falsity of Christianity, Ruse, a nonbeliever, wrote *Can a Darwinian Be a Christian?* He looks at every imaginable point of contact between evolution and Christianity and answers yes to the question posed in his title:

> If you are a Darwinian or a Christian or both, remember that we are mere humans and not God. We are middle-range primates with the adaptations to get down out of the trees, and to live on the plains in social groups. We do not have powers which will necessarily allow us to peer into the ultimate mysteries. If nothing else, these reflections should give us a little modesty about what we can and cannot know, and a little humility before the unknown.[16]

LOOKING AHEAD: THE PLAN OF THE BOOK

The creation–evolution controversy in America has become so overheated and loaded with half-truths and nonsense that it is all but impossible to get a clear picture of anything. Mythologies abound on both sides. Darwin's apocryphal deathbed repudiation of evolution is a popular and widely circulated myth comforting the faithful. The imminent collapse of evolutionary theory and the occasional celebrated negative comment about evolution by a leading scientist are others. These offer hope that biblical creation will make a comeback in America. Mirror-image mythologies about evolution are equally plentiful: the theory provides a solid foundation for atheism and assures the ultimate victory of secularization; every intelligent person now believes it; dissenters are "stupid, wicked, or insane."[17] We even hear that evolution will soon explain religion away. Such affirmations assure blinkered secularists that someday religion will go extinct, eaten away by the acid of evolution.

In the pages that follow I offer readers a tour of this troubled battlefield. Darwin, we will see, began his career as a committed Christian. He planned to become a minister and certainly had no intention of undermining religion. That his theory did this kept his stomach in a constant knot. Nevertheless, the responses to his theory, even from religious conservatives, were not uniformly hostile and, almost immediately, thinkers were finding ways to incorporate this new view of origins into their theological understanding of creation. Some even welcomed the theory as a more satisfactory explanation for nature's excessive waste and carnage. The widespread hostility currently leveled at Darwin's theory is a recent development, although it has always been present to some degree.

The most interesting and often unintentionally humorous challenges to Darwinism have not been scientific, but legal. Curiously, a cavalcade of lawyers claiming to have detected logical flaws in evolutionary reasoning starts with one of Darwin's contemporaries and runs through to some prominent lawyers in the present. Some of these lawyers, strangely, actually boast of their ignorance of biology as they flail about in irrelevance.

Moving into the present we encounter "scientific creationism" (also called "creation science") and "intelligent design," sibling perspectives insisting they are unrelated. Despite being largely devoid of scientific content, these movements have captured the hearts and minds of over half the country, although they remain, for the time being at least, banned from America's public schools.

In the current controversy, science has disappeared, and the argument has turned into a culture war, with political allies in smoke-filled back rooms formulating strategies with little regard for truth. Meanwhile, off the front pages of the newspapers, the science of evolution grows increasingly robust and secure, even as America's schools find the topic increasingly harder to teach.

I wish I could promise that the story in the following pages has a happy ending, but it does not. Loud confident voices, including the echo of my own college worldview, assure us that evolution is a false theory being used by Satan to destroy faith in God; equally loud voices counter that evolution is a true theory that is destroying faith in God. Quiet but less confident voices point out the absurdity of both of these claims. This disagreement is not going away anytime soon.

Places exist on which believers can stand, however, in the midst of the controversy. We don't know anywhere near enough about evolution to infer from it that God is not the creator. And we don't know anywhere near enough about God to dismiss the idea that evolution might be a part of God's creative processes. If we can embrace a bit of humility and avoid the temptation to enlarge either evolution or biblical literalism into an entire worldview, we can dismiss this controversy as the irrelevant shouting match that it is.

These insights, of course, were nowhere in sight as I began to wrestle in college with the unwelcome truth that evolution had strong empirical support and could not be dismissed as a satanic delusion. As I look back after three decades of reflection I can see, however, that my sophomoric struggles were nothing more than my personal encounter with Darwin's dangerous idea, an encounter that was hardly original with me. Believers everywhere, especially in America, continue the search for the elusive role that evolution should play in a comprehensive and satisfying understanding of ourselves and our origins.

CHAPTER 1

THE LIE AMONG US

H istory records three Charles Darwins. The most interesting Darwin is the one who repudiated his theory of evolution on his deathbed. A colorful character named Lady Hope claimed to have visited Darwin on his deathbed, where she found him reading his Bible and recanting his life's work. "I was a young man with unformed ideas," she quotes him as saying. "I threw out queries, suggestions, wondering all the time over everything. And to my astonishment the ideas took like wildfire. People made a religion of them."[1] Lady Hope's winsome story, which historians have shown was a complete fabrication,[2] has been circulating broadly among American evangelicals for the better part of a century and can still be found there.[3]

History's second Darwin is a sinister character in a story even more popular among evangelicals than Lady Hope's fiction. This Darwin was an enthusiastic and committed unbeliever who combed the globe gathering evidence to rationalize his disbelief. Authors and television personalities John Ankerberg and John Weldon present this Darwin in their popular *Darwin's Leap of Faith*. They argue that Darwin himself never even found evolution convincing. Their demonized Darwin rationalized atheism by concocting a preposterous theory whose only saving grace was its demolition of the idea that God created the world. To "soothe his fears," Ankerberg and Weldon write, "Darwin adopted a philosophy convenient to his own rejection of God."[4] This Darwin is also a fabrication, although less entertaining than the Lady Hope myth. Reading any one of the many recent excellent biographies of Darwin will put this to rest.

The third and actual Darwin was neither a deathbed convert nor lifelong crusader against belief in God. He was, in fact, a sincere religious believer who began his career with a strong faith in the Bible and plans to become an Anglican clergyman. He did eventually lose his childhood faith, but it was reluctantly and not until middle age, long after his famous voyage on the *Beagle*. Toward the end of his life he wrote to an old friend about the painful experience of losing his faith: "I was very unwilling to give up my belief." He recalled daydreaming about something that could arrest his slide into disbelief, perhaps the discovery of "old letters between distinguished Romans, and manuscripts being discovered at Pompeii or elsewhere, which confirmed in the most striking manner all that was written in the Gospels." Gradually, though, he found it harder to imagine being rescued in this way, and "disbelief crept over me at a very slow rate, but was at last complete."[5]

THE DEMONIZED DARWIN

Unfortunately, the real Darwin is the only one of no interest to anti-evolutionary demagogues. Eager to keep the faithful on track, they smear Darwin and his theory unmercifully. In *The Long War Against God: The History and Impact of the Creation/Evolution Controversy*, the late Henry Morris proposed that Darwin actually got his theory indirectly from Satan. Darwin, argues Morris with a perfectly straight face, was simply one in a long line of dupes spreading a sinister gospel of materialism originally delivered to humanity by Satan at the Tower of Babel.

> The very first evolutionist was not Charles Darwin or Lucretius or Thales or Nimrod, but Satan himself. He has not only deceived the whole world with the monstrous lie of evolution but has deceived himself most of all. He still thinks he can defeat God because, like modern "scientific" evolutionists, he refuses to believe that God is really God.[6]

Ken Ham, who heads the popular Answers in Genesis organization and is currently America's leading creationist, sees an apocalyptic dimension to evolution. On the back cover of his book *The Lie: Evolution,* Ham writes: "The Bible prophetically warns that in the last days false teachers will introduce lies among the people. Their purpose is to bring God's Truth into disrepute and to exploit Believers by telling them made-up and imagined stories. Such a Lie is among us. That Lie is Evolution."[7]

And finally, in more careful, restrained, and intentionally secular-sounding prose Phillip Johnson, the leader of the intelligent design movement, says:

"The aim of historical scientists—those who attempt to trace cosmic history from the big bang or before to the present—is to provide a complete naturalistic picture of reality. This enterprise is defined by its determination to push God out of reality."[8] In Johnson's opinion natural science is far too natural.

To the amazement of most Europeans, who made their peace with evolution long ago, these views on Darwin and his theory are widespread in the United States. The majority of children raised in America's evangelical culture encounter them somewhere, often from creationist evangelists like Ham who head organizations dedicated to destroying evolution. Ham's Answers in Genesis organization, for example, has almost two hundred employees and sponsors thousands of events every year, from visits to churches to massive rallies in public arenas with music and multimedia presentations. Sixty thousand people visit his Web site every day, and his books, videos, and tracts sell well. A $27 million creation museum opened in 2007. A glossy magazine, *Answers,* goes out to almost fifty thousand readers.[9] And, although Ham's operation is the most polished and best funded, there are dozens of others like it. Spreading the gospel of anti-evolution, with Darwin as the villain, is a million-dollar industry reaching an eager audience of American evangelicals larger than the population of any country in Europe.

But Charles Darwin is not a villain, and these portraits of him are irresponsible and malicious caricatures distorting him beyond recognition. In their eagerness to turn Darwin into a scary boogeyman, his detractors rewrite history and invent motives to suggest that evolution began as a conspiracy to destroy belief in God.

THE TORMENTED EVOLUTIONIST

Charles Darwin was born in 1809 to a well-to-do British family who, despite having some unorthodox characters listed in the family Bible, raised him in the Anglican Church, educated him at an Anglican school, and put him on the train to Edinburgh to study medicine. When this career ran off the rails, Charles's father, fearing his son might become an "idle sporting man," sent him to Cambridge to study theology in the hopes that he might become a parish priest.[10] Charles obliged, for he took family obligations seriously and was attracted to the genteel life of a "country clergyman." Nevertheless, he did look closely at the affirmations of the Anglican creeds, but since he did not "in the least doubt the strict and literal truth of every word in the Bible," he concluded that the creeds were acceptable, if confusing.[11] Whatever radical genes the young Darwin may have possessed had not yet kicked in.

Darwin's interest in natural history was enriched through his study of both medicine and theology, which were nicely complementary pursuits in nineteenth-century England. Science—which medicine aspired to be—nestled within a framework of natural theology, which uses insights from nature to fashion arguments about God. The most common argument was the traditional claim that design in nature implied the existence of an intelligent creator. Unlike today, when theology and science reside in different buildings on opposite corners of university campuses separated by armed guards and barbed wire, at this time they were in a robust and congenial dialog. Many parish priests were active naturalists, and there was a consensus that the rapidly developing sciences would continue to provide useful theological insights.

While studying theology at Cambridge University, Darwin came under the spell of William Paley, a leading Anglican philosopher and passionate abolitionist. Paley's influential texts, *Natural Theology, The Principles of Moral and Political Philosophy*, and *Evidences of Christianity*,[12] were standard fare for students of Darwin's generation and greatly influenced nineteenth-century British thought. Darwin would later comment that he could probably "have written out the whole of the *Evidences* with perfect correctness."[13] The design in nature, articulated Paley with arguments so clear and compelling they were compared to those of Euclid, implied the existence of a designer, namely, God. Darwin and his generation were taught to see the handiwork of God in nature; its beauty, order, and rich creativity reflected the attributes of its creator.

Darwin's career took a critical turn when Captain Robert Fitzroy, a conservative Anglican, accepted a recommendation that Darwin join him on an epic journey around the globe on a modest ship named the *Beagle*. The primary agenda was a survey of South America, though Fitzroy also intended to return some Fuegians who had trained in England as missionaries to Tierra del Fuego.

Extended journeys at sea were often lonely affairs for captains, typically the only cultured member of a tiny community of illiterate, seafaring philistines, all of them male and living in close quarters for months on end.

Fitzroy's uncle had committed suicide at sea, probably driven to it by intense loneliness. The *Beagle*'s captain needed a companion with whom to eat, talk, and stay sane. The passenger could also function as the ship's naturalist, cataloging the exotic flora and fauna of the globe to the greater glory of God and Great Britain.

Enter twenty-two-year-old Charles Darwin, for whom this posting was custom-made. After some negotiations with his father, Darwin joined the crew of the *Beagle* and set sail from Plymouth harbor two days after Christmas in 1831. The *Beagle* would return five years later with her captain still sane and her famous passenger in a muddle.

Darwin boarded the *Beagle* with his childhood Christian faith intact, although he had begun to wonder about the historicity of the more fanciful Old Testament stories, like the Tower of Babel. He was also starting to wonder about the vengeful, tyrannical God of the ancient Israelites. At one point during the voyage he recalled being "heartily laughed at by several of the officers for quoting the Bible as an unanswerable authority on some point of morality."[14] For the most part, however, Darwin's faith was unruffled, with the exception of his natural theology, which was constantly pierced by troubling observations that defied his expectations.

UNNATURAL THEOLOGY

Naturalists of Darwin's generation, like most scientists before and since, studied nature within the framework of their best understanding of the natural world. It is a popular fallacy that scientists study nature with no expectations, their observations falling on mental blank slates to be organized with perfect objectivity into secure and dispassionate generalizations that do nothing more than summarize the facts. Observations, rather, are gathered to *test* various ideas that are in play. Most often the ideas pass the tests and become more secure as a result, but sometimes the observations raise important questions. Darwin, like all scientists, brought his *expectations* to his observations of the natural world, constantly checking to see if the new facts were *consistent* with the expectations.

The network of expectations guiding scientific research at any given time is called a *paradigm*. It represents the collective wisdom of the scientific community and would have been reflected in the textbooks and lectures that

Darwin encountered at the university. Science advances, in general, by refining the understanding of these paradigms and by bringing more and more observations under the paradigm's explanatory umbrella.[15]

The role played by paradigms in science is paradoxical and can appear suspicious to outsiders. How can it be that a scientist like Darwin can *start* his investigation of nature "with his mind made up," so to speak? Are not these assumptions equivalent to *prejudices* blinding scientists to the truth and preventing them from correctly interpreting their observations? Does this not turn scientific investigation into a simple rationalization of the status quo? Was the Darwin of the *Beagle* simply reading his preconceptions into his observations of nature?

These questions are entirely legitimate. Nevertheless, there is a simple response: this is how the science that cured smallpox, built the atomic bomb, and put a man on the moon works. Centuries of rapid and creative scientific advance have honed the methods of science to the point where most people simply have faith in science. Advertisers exploit this faith when they describe their claims as "scientific" facts, as if some facts are more factual than others. Most people today are quite content to check into a hospital and place their very lives in the hands—one is tempted to say "on the altars"—of this science.

Under normal circumstances paradigms offer helpful guidance. In the century before Darwin, for example, Newton's law of universal gravity was an important guide to understanding the motions of a growing roster of celestial objects. After a few spectacular successes, astronomers stopped wondering if the law was correct. They simply assumed that it was and saw their assumption repeatedly validated.

Paradigms become interesting when they start to fail, which was what Darwin experienced on the *Beagle*. Long-standing assumptions about the natural world, buttressed by the authority of countless experts and integrated into comprehensive visions of reality, are challenged by fresh observations. "Commonsense" views of the world begin to crumble; order descends into chaos and understanding into confusion. These radical, world-shaking developments receive the label "scientific revolutions."

Scientists typically embrace their paradigms with a tenacity bordering on the irrational. In my training as a physicist I was simply taught the laws of physics, with no hint that they were anything other than decrees handed down by God to people like Albert Einstein and Niels Bohr. When it came time to start laboratory work, I did not consider for a moment that my experiments might contradict what I had learned; if I had run into my adviser's office waving a graph I said refuted quantum theory, my adviser would have laughed hysterically and suggested I switch to philosophy. Science is an incredibly conservative enterprise. Nevertheless, practicing scientists are anything but conservative and are quite often eccentric iconoclasts, or "nerds" in popular parlance. Dreams of revolution inspire scientists—of being the next Einstein and laying waste to the status quo—but the great staying power of their paradigms keeps them on course.

The inertia of paradigms is, paradoxically, the very reason we can trust science. New ideas in science are subjected to a withering scrutiny before they are accepted. Old ideas must be thoroughly refuted before they are discarded. If a long-standing and traditional idea, like astrology or a young earth, has been abandoned by science, we can be confident that it was not without compelling reasons.

So how do scientific revolutions occur? They start with observations that don't fit. Initially these "observations" often don't even register, like parents who can't see that their son is a bully. Then they register and become puzzles of great significance that raise questions about the prevailing paradigm. (Why does my son have no friends?) And then they become ho-hum facts that fit into a new paradigm. (My son is a bully, and his peers don't like him!) A classic example from the history of astronomy reveals this pattern.

STAR LIGHT, STAR BRIGHT, THE STAR I CAN'T SEE TONIGHT

In 1054 a brilliant new star appeared brightly in the constellation Taurus. Four times brighter than Venus, it was visible in daylight for several weeks and at night for almost two years. Enthralled Chinese astronomers wrote extensively about it and what it meant for developments on earth. European and Arab astronomers, however, were strangely silent about the new star, as if they did not even see it. They did "see" it, of course, for they were active observers, and the star was not to be missed. The only explanation for their oversight is that a new star was so thoroughly inconsistent with their expectations that they could not accept the testimony of their own eyes. Their astronomical paradigm included an ancient belief that the heavens were unchanging. The appearance of a new star would entail the absurd proposition that God had "restarted" the process of creation that had been completed on the sixth day. And that, of course, was simply ridiculous.

When such "impossible" observations become accepted as real, they "register" and raise deep questions about the veracity of the reigning paradigm. In 1572 Europe's greatest astronomer, Tycho Brahe, observed another new star in the heavens and was dumbfounded:

> Amazed, and as if astonished and stupefied, I stood still, gazing ... intently upon it.... When I had satisfied myself that no star of that kind had ever shone forth before, I was led into such perplexity by the unbelievability of the thing that I began to doubt the faith of my own eyes.... And at length, having confirmed that my vision was not deceiving me, but in fact that an unusual star existed there ... immediately I got ready my instrument. I began to measure its situation and distance from the neighboring stars.[16]

Brahe's new star made no sense within the reigning explanatory paradigm. But there it was, visible in daylight, clearly a new star. In time such observations helped topple the reigning paradigm, and astronomers became comfortable with the idea that new things occasionally appeared in the heavens.

In 1987 another new "star"—Supernova 1987a—appeared and, although it made the cover of *Time* magazine, it occasioned no distress in the scientific community.[17] By 1987 such phenomena had become part of mainstream astronomy, and this time the new star was of interest largely because it confirmed some untested implications of theories of stellar evolution.

DARWIN AND THE PARADIGM OF INTELLIGENT DESIGN

Darwin's observations on the *Beagle* mimic those of astronomers reacting to the appearance of new stars that don't fit into accepted paradigms. When he boarded the *Beagle,* Darwin had a traditional Christian worldview. On a personal level, he trusted the Bible and looked to it for moral guidance. On a philosophical level, he believed strongly in God as the source of the created order and the foundation for belief in salvation and eternal life. And, on a scientific level, he believed that the natural world was intelligently designed and that its design spoke clearly and eloquently of the wisdom, love, and creative power of God. The Darwin of the *Beagle,* like all naturalists of his generation, looked at the world through the same eyes as the contemporary proponents of intelligent design, who see the handiwork of God in nature's intricate machinery.

Paley and the other natural theologians shaping Darwin's era had created a compelling framework for understanding the world as a collection of elegantly designed organisms flourishing in custom-made ecological niches. Hydrodynamically sophisticated fish swam in water, and aerodynamically sophisticated birds flew in air. Like toddlers on a playground with brightly colored and unusually safe equipment that is "just their size," the flora and fauna of planet earth flourish in environments designed for them. The world and its inhabitants were, quite literally, *made* for each other, and everything everywhere testified to the glory of God.

It would be hard to overstate the importance of divine design among British naturalists of Darwin's generation. The patterns of nature were all attributed to God; the roster of living creatures was organized in a "great chain of being" that revealed the hierarchical structure of the created order, progressing from simple to complex; the properties of water and air and soil and weather reflected God's wisdom and care.

Paley's *Natural Theology,* published in 1802, was one of the most popular texts in the English language. Read by all of Britain's naturalists, it provided the paradigm for understanding the natural world. As Darwin gazed over the railing of the *Beagle,* he saw the world through spectacles provided for him by William Paley. The handiwork of God was everywhere visible. That he was often leaning over the rail being sick did nothing to dissuade him from his conviction that the world, including his own troubled digestive system, was a grand machine crafted by the Great Mechanic.

NATURAL THEOLOGY

Creationists have launched a salvo of accusations at Darwin, claiming he invented evolutionary theory to rationalize his lack of faith. These claims are so blatantly false and so clearly in opposition to everything we know about Darwin, that we have to wonder how they arose. The facts are quite clear: Darwin inherited a worldview that was solidly creationist, although that term was not in use at that time. The young Darwin could have been a staff biologist at Henry Morris's Institute for Creation Research or perhaps a tour guide in Ken Ham's creation museum. Certainly he could have been a senior fellow at the Discovery Institute, helping Phillip Johnson write op-eds and popular books promoting intelligent design.

Darwin, as we know, eventually abandoned this way of looking at the world. But this transition did not derive from his creeping agnosticism. It resulted from his repeated discoveries that the world was full of things that did not look intelligently designed. Eventually he slowly, and quite reluctantly, began to wonder whether there might be a better explanation for the observations that were his passion.

The young Darwin was, in fact, the equivalent of today's "intelligent design theorist," and perhaps it is as a traitor to this viewpoint that he generates so much hostility from his twenty-first-century counterparts. But there is no historical ambiguity about the central role that ideas about intelligent design played in his thought.

Darwin worked within an intellectual tradition that had been doing science—then called *natural philosophy*—in a theological context for centuries. And, although the term *intelligent design* was not in use at the time, there is little difference between this tradition and what currently bears the label. The only real difference is *political:* contemporary intelligent design is at war with mainstream science, while its precursor was in harmony with science. The drama of Darwin's generation was further reduced by the almost complete absence of polemicists like Richard Dawkins using science as a weapon against religion. Virtually the entire scientific tradition from Galileo to Darwin was deeply religious.

Science before Darwin rarely ran afoul of religion.[18] Galileo is a notable exception, but his conflict with the Roman Catholic Church represents only one aspect of his rich and varied scientific career. Most interactions were less exciting, and many were actually constructive. Not long after Galileo, for example, Newton discovered that the universe ran by a few simple laws. This led to the idea that the universe was like a great clock, implying that there must be a Grand Clockmaker who created it.

In the near perfect circular motion of the planets around the sun, Newton discovered an astonishing balance between the speeds of the planets and the stability of their orbits. Slow them down and they spiral into the sun; speed them up and they spiral away and leave the sun's gravitational embrace. This delicate balance was but one of many impressive features of the Newtonian world machine, exhibiting what mathematicians call beauty. Everywhere Newton looked he saw clear evidence of design. In his most important work, the *Principia Mathematica,* first published in 1687, Newton wrote: "This most beautiful system of the sun, planets, and comets could only proceed from the counsel and dominion of an intelligent and powerful Being." He goes on to mention other examples of God's wisdom, like the placement of the stars at great distances from each other, "lest the systems of the fixed stars should, by their gravity, fall on each other."[19]

Newton lit a fuse that ignited an explosion of scientific knowledge that transformed the following centuries. He modeled a way of using science to support religious belief, an approach rooted in the Middle Ages, when science was known as the "handmaiden" of theology. It was an approach that would carry forward into Darwin's century and even into our own, although with modification.

Across Europe, amid the splintering Christian denominations and even in the emerging deism, science supported natural theology. The marvelous hand of God was readily discerned in creation, as Newton, Paley, and everyone in between made so clear. Scientists were fascinated by a range of mysteries, including the enigmatic character of everyday occurrences like ice. Mysteriously, it is less dense than water and floats, enabling creatures to survive beneath it, protected from the ravages of cold northern winters. Does this not reveal the wisdom of God? Consider the eye. How could so many intricate parts—balls, sockets, lids, lenses, retinas, optic nerves—come together and work so well? Human joints, bats' wings, mother's milk, chicken eggs, roots, leaves, wind, rain—all celebrated the glory of God. Books with titles like *Water Theology* and *Insect Theology* argued directly from the details of creation to the nature and existence of the creator. God's fingerprints were everywhere.

Even those starting to reject Christianity and the Bible found in nature a compelling witness to God as creator. Thomas Paine, who penned the notorious *Age of Reason,* in which he claimed to "detest" the Bible "as I detest everything that is cruel," found in nature a clear revelation of God's power and benevolence.[20] The Bible, Paine contested, was written by men; God wrote the book of nature. The Bible was parochial and recent; nature was ancient and universal, available to all people at all times. Such celebrations of nature were common across Europe and in the New World. Everywhere, science supported belief in God through its revelations of both God's wisdom and concern for creatures. This tradition of natural theology nurtured the young Charles Darwin who set sail on the *Beagle.*

To be sure, there were exceptions. The Scottish skeptic David Hume, for example, challenged any argument claiming to identify divine "design." Perhaps, he suggested, design in nature was illusory or unintended. Glasses sit neatly on one's nose, but who would argue that the nose was made for this purpose? And some design looks stupid, even malevolent. Consider, Hume wrote, the many "curious artifices of nature, in order to embitter the life of every living being."[21] The French satirist Voltaire lampooned the idea that the world was well designed for its inhabitants. Appalled by the Lisbon earthquake, which killed a hundred thousand people, Voltaire ridiculed the popular idea that this was the "best of all possible worlds."[22]

Dissenters like Voltaire, who continued to believe in God, and Hume, who did not, could not hear nature testifying to a wise and benevolent creator. But they were minority voices, remembered as cranky renegades at odds with more traditional notions tucked deep into the hearts of their fellow Europeans. Such naysayers did little to chase natural theology from Britain, where earthquakes of the sort that destroyed Lisbon had never disrupted the blessed and bucolic countryside.

The centerpiece of nineteenth-century natural theology, of course, was William Paley's 1802 classic *Natural Theology, or Evidences of the Existence and Attributes of the Deity Collected from the Appearances of Nature,* where we find his famous watchmaker analogy:

> In crossing a heath, suppose I pitched my foot against a stone, and were asked how the stone came to be there; I might possibly answer, that, for anything I knew to the contrary, it had lain there forever.... But suppose I had found a watch upon the ground, and it should be inquired how the watch happened to be in that place; I should hardly think of the answer I had before given.... There must have existed, at some time, and at some place or other, an artificer or artificers, who formed [the watch] for the purpose which we find it actually to answer; who comprehended its construction, and designed its use.... Every indication of contrivance, every manifestation of design, which existed in the watch, exists in the works of nature; with the difference, on the side of nature, of being greater or more, and that in a degree which exceeds all computation.[23]

Paley's watchmaker analogy—a standard part of the early nineteenth-century curriculum in England—bears exactly the same form as arguments that would be made two centuries later by intelligent design proponents. Compare this passage by the organizers of a major intelligent design conference two centuries after Paley:

> The universe and its laws have not always been around in their present state. The data from science also suggest a high degree of complexity throughout the history of life, and such complexity requires explanation that not only includes but also transcends natural processes alone. In addition, the data from science indicates an incredibly high degree of fine-tuning or balance within the structure of the universe at all levels. This also calls for an explanation that transcends natural processes.[24]

This is a critically important part of our story, as it illustrates the vitality of intelligent design thinking at the time of Darwin and, I am arguing, makes the early Darwin a nineteenth-century intelligent design theorist.

Paley's book was an eloquent summary of a broad range of arguments that had been developed over the preceding decades. Here we have an ingeniously fashioned wing; there a clever fin; look at this eye; consider this antenna; marvel at this or that appendage with this or that specific function. In compelling and captivating prose, the prose Darwin could quote by heart, he summarized an impressive range of design in the natural world and how this design pointed with clarity to the existence of a designer, "an intelligent designing mind for the contriving and determining of the forms which organized bodies bear."[25]

As Darwin boarded the *Beagle* the design of the natural world was as clear to him as the design of the boat that would carry him around the planet. Both were obviously the work of intelligent designers who matched form to function, and the *Beagle* lived up to expectations. The natural world, however, repeatedly failed to match Darwin's expectations. Each time the *Beagle* put down its anchor and Darwin inspected the local flora and fauna, he returned with troubling questions. Trained to believe that the natural world revealed a benevolent and wise creator, he began to wonder why so much of the world looked neither wise nor benevolent.

TROUBLING QUESTIONS

Scientific revolutions are three-act plays. In the first act, the status quo is so universally accepted that people have trouble even noticing ill-fitting anomalies. A new star where there is not supposed to be one will be over-looked or dismissed as irrelevant. In the second act, anomalies are noticed but viewed as puzzles to be solved, it is hoped, within the framework of the status quo. The new star creates a crisis forcing examination of the prevailing framework to see if it can be adjusted to accommodate this irregularity. In the final act, the anomalies precipitate the collapse of the status quo and become evidence supporting an entirely new understanding. Here is a "new" star, and it makes perfect sense.

Darwin's thought followed this same trajectory. He started his career as a naturalist viewing the world through the lens of natural theology and see-ing intelligent design. But then he began to notice things that didn't fit: here is an animal with webbed feet living on dry land; there is a bee that dies after stinging its prey, its stinger serrated in a way that prevents extraction after in-sertion; here is a cat apparently torturing a mouse before killing it.

To suggest that these examples manifested God's wisdom and benevo-lence made a mockery of those terms. Did the loving God of Darwin's youth *really* install instincts in cats that would make them enjoy pummeling mice as if they were feline loan sharks from a barnyard parody of an old gangster movie? Surely not. Like Brahe observing a new star, Darwin made observa-tions that challenged the bedrock assumptions of his paradigm.

The *Beagle* was a small ship, some twenty-four feet wide and ninety feet long. Cramped quarters provided limited room to maneuver, adding to the stress of the long journey. The framework of natural theology within which Darwin worked was similarly cramped and offered little room for intellectual maneuver. The anomalies that bothered Darwin had responses, of course. Maybe we just don't see the big picture; perhaps sin and the fall are respon-sible for some of the problems; maybe we don't understand the phenomena well enough; and so on. But these responses are woefully inadequate and little more than patches on an ancient ship riddled with holes and taking on water.

BATS, CATS, AND WASPS

As befits one of our species' true revolutionaries, scholars have scrutinized Darwin in detail. Every scribble in his voluminous notebooks and every let-ter in his vast correspondence have been dissected; every scientific paper has been examined for hints of the revolution to come; every footnote is a possible shaping influence, every acquaintance a possible intellectual accomplice. His modest autobiography, written near the end of his life and based on fragile recollections, has been laid out beside his more historical notebooks and the discrepancies analyzed. Magisterial new biographies appear with regularity, each one updating our unfolding picture of the nineteenth century's greatest scientist. The result is a clear picture of how Darwin came to his theory.

The natural theology of Darwin's training explained the distribution of life on the planet as God's coordinated design of both creatures and their habitats, an explanation that accounted for the many remarkable adaptations. But some things didn't fit. In South America, to take one example, Darwin encountered a new species of rhea, a flightless bird living on the pampas of Patagonia in an area adjacent to that of the common rhea. Each species of rhea had its own territory, but there was a large contested area between them that they shared. The rhea posed puzzles. The most obvious was the idea of a flightless bird. Why would God create a bird with so much unused aerodynamic paraphernalia? Why would God place two virtually identical birds in different habitats? And, finally, what was up with the pointless competition between the two species for control of the borderlands separating them? The humble rhea embodied a set of contradictions that even Paley would have had trouble rationalizing as the handiwork of God.

Similar difficulties cropped up all over the planet. Darwin noted an upland goose that never went in the water, yet was handicapped by webbed feet. If this was the handiwork of God, it was surely a cruel joke, as anyone who has ever tried to walk in flippers knows only too well. There were birds resembling woodpeckers with all the necessary facial reinforcements to pound their heads constantly against a tree, and yet they lived on insects found on the ground. God seemed to be wasting resources in giving these birds such overdesigned beaks.

The geographical distribution of animals puzzled Darwin. Charles Lyell offered one explanation in his influential *Principles of Geology*,[26] which Darwin was reading carefully while aboard the *Beagle*. Lyell was among the emerging "scientific geologists" working to free their new science from "Mosaic geology," which they regarded as "marginal" and "worthy only of derision."[27] These geologists, on scientific grounds, rejected the flood of Noah and its implication that the worldwide distribution of animals derived from their dispersal from Mt. Ararat, where the ark came to rest. Alternative explanations for the distribution of animals still invoked divine creation, of course, but in ways based on empirical, rather than biblical, considerations. In Lyell's view, with which Darwin would wrestle, God had placed individual species in "centers of creation" specifically prepared for them. God created the earth with its various habitats—deserts, meadows, swamps, mountains, rivers, oceans, islands, cold climates, hot climates, and so on—and then created animals to flourish in the different habitats. Darwin thus anticipated that animals indigenous to these centers of creation would have features optimized to the local conditions.

Contrary to expectations, however, Darwin could not explain the distribution of animals he encountered. Why, for example, were certain islands populated by bats but no other mammals of any sort, when they would have provided wonderful habitats for many mammals? Was it just a coincidence that the only mammal on these islands was one that could have flown there on its own? Why did each of the Galapagos Islands have its own species of tortoise, so easily distinguished that the locals could simply look at a tortoise and tell you the island from which it came? If God matched species to their habitats in centers of creation, as Lyell believed, why would identical habitats have different species?

None of these observations ruled out the possibility that God was still the creator of all the life-forms on the earth. But they did raise troubling questions about the mechanisms of creation and the degree to which God was involved in the details. Darwin described such phenomena as "utterly inexplicable on the theory of independent acts of creation."[28] This is the first level of Darwin's concern—the intelligent design paradigm could not explain many of the details of the natural world.

An even stronger conviction that God was not responsible for the details came from Darwin's growing awareness of natural phenomena so horrible it was inconceivable that they embodied plans originating in the mind of God. For example, the way Ichneumonidae wasps feed off the internal organs of their caterpillar hosts appalled Darwin. The mother wasp inserts a paralyzing chemical into the nervous system of the caterpillar and then places her eggs inside the still-living host, where they hatch and then gradually devour the paralyzed caterpillar from the inside. The hatched baby wasps emerge with preprogrammed instincts to consume the internal organs of the caterpillar in a sequence that keeps their caterpillar host alive as long as possible.

Such examples posed disturbing challenges to natural theology. The system by which Ichneumonidae eggs hatch is truly ingenious, although the host caterpillars might prefer a different term. Variations on the theme show up regularly in science fiction movies about aliens that parasitize human hosts. In the classic *Alien* films, aggressive alien parasites take over human bodies by attaching to their faces, inserting tubes down their throats, and planting embryos inside them. When the embryos are mature, they explode out through the chests of their human hosts, killing them and scaring the bejeezus out of the audience.

Nonfictional horror shows like the creepy Ichneumonidae and sadistic cats bothered Darwin. How were they to be reconciled with his belief in creation? On the living-room floor a kitten is entertaining as it plays with a ball of yarn, and it would be easy to see this as simply delightful. But outside in the yard, the kitten's mother, influenced by the same instincts, is beating up a mouse that she may or may not eat after she kills it.

In a letter to the American biologist Asa Gray in 1860, a year after he had published *On the Origin of Species* and twenty-four years after getting off the *Beagle,* Darwin was still wrestling with these issues: "I cannot see, as plainly as others do," he wrote, "evidence of design and beneficence on all sides of us. There seems to be too much misery in the world. I cannot persuade myself that a beneficent and omnipotent God would have designedly created the Ichneumonidae with the express intention of their feeding within the living bodies of caterpillars, or that a cat should play with mice."[29]

ON THE ORIGIN OF SPECIES

Revolutionary ideas in science rarely come roaring down the track with belching smoke, piercing whistles, and squealing brakes. They arrive more like a gathering storm. A cloud appears, here and there, in a blue sky. A drop of rain is felt. More clouds. More rain. The sky becomes partially, then fully, obscured. The sun is blotted out. A bit of thunder and lightning creates drama, the clouds begin to break, and the sun reappears. But it is not the same sun, and everything looks somehow different.

Darwin grew dissatisfied with the prevailing creationist ideas. They made no sense theologically, and they offered almost nothing scientifically. His growing dissatisfaction was a gathering storm; puzzles like the Ichneumonidae, cats, rheas, and flightless birds were its clouds. Eventually the old sun was blotted out, and the landscape became hard to see and impossible to comprehend. It was in the rain and fog of this storm that Darwin developed his theory of evolution.

Darwin circulated his theory privately among close friends for two decades before publishing, nervous about the anticipated controversy. But eventually, prodded by the awareness that a fellow naturalist, Alfred Wallace, had developed an identical theory, Darwin published what turned out to be a most paradoxical theory—one that combined great explanatory power and theoretical simplicity. "Why didn't I think of that?" responded many of his associates.

His theory is disarmingly simple. Darwin begins by noting the great competition in nature. Most species produce far more offspring than can survive. As a child I loved to gather the little "helicopters" dropped by the mighty maple in my yard. One maple can drop up to seven thousand of these twirly seeds in a single year, enough to create a large forest, if they were all to survive. Similarly, one spawning salmon can release five thousand eggs each year, enough to stock a lake.

However, most attempts at reproduction fail. No salmon has five thousand babies that grow to maturity. The ones that succeed, argued Darwin, do so because they are more fit, better able to meet the challenges of the local environment. This enhanced fitness can be passed on to the next generation. In this way, species evolve slowly, imperceptibly, as they become better adapted to their local environments. Fish grow ever more hydrodynamic; hawks get better vision; camels store larger quantities of water.

Sometimes, however, the local environment changes. A river dries up, a peninsula breaks off into an island, a new predator arrives, an earthquake moves a beach up on to the side of a slope, an avalanche covers the mouth of a watery cave. Such changes alter the environment, and previously well-adapted species face new challenges. A goose with webbed feet that evolved to accommodate swimming may be relocated away from the water. Turtles confined to a newly isolated island will evolve independently of their siblings on the mainland. Birds with powerful beaks may no longer find prey in trees. And so on.

Such modifications to the environment pose new challenges. Take the goose with webbed feet, now constrained to make its way on dry land. The webbing between its toes, once useful for moving in the water, is now an encumbrance, making walking slow and awkward. What was useful in one environment is a disadvantage in another. And, although natural selection may gradually minimize the problem of webbed feet, there is no mechanism available to simply remove it. Natural selection tinkers with existing traits relevant to reproduction, making them ever more useful in the existing environment. Natural selection, however, cannot suddenly make wholesale changes or undo developments long in the making.

For Darwin, explanations like these made more sense than supposing that God had placed a goose with webbed feet on dry land or that the goose had walked there after disembarking from Noah's ark. Darwin's explanations illuminated countless oddities across the globe that made no sense within the explanatory paradigm of intelligent design.

Natural selection, operating on tiny changes in organisms over vast periods of time, accounted for much of what Darwin was struggling to understand. Even the Ichneumonidae were less disturbing when viewed as the product of natural selection rather than the direct handiwork of God. By these lights, God no longer seemed like a cruel despot, creating monsters to prey on innocent life; the villain doing the dirty work was now a blind and impersonal process of natural selection. Darwin found this interpretation far more congenial than the theological gymnastics required to fit nature's monstrosities into Paley's framework of natural theology.

Although Darwin rejected the idea that God was responsible for each individual organism, he continued to believe that God played a role in nature. In the same 1860 letter to Asa Gray expressing his disgust at the Ichneumonidae, he noted that he could not be "contented to view this wonderful universe and especially the nature of man, and to conclude that everything is the result of brute force." He preferred instead "to look at everything as resulting from designed laws."[30] God, he suggested, may have created the vast physical framework in which natural history unfolded, charting its own course, sometimes for better and sometimes for worse.

Darwin's critics write as if this suggestion—that nature has its own freedom within a framework of laws designed by God—is an appalling and antireligious stance. This is an odd response. Christian theology has always had a place for freedom, even for the followers of John Calvin, with their predestination; even they can smuggle in a bit of free will for themselves. Christian theology embraces the very human freedom to create or destroy, to choose evil or good, to promote life or death. Darwin's invocation of chance in nature is equivalent to granting the natural order some measure of the very freedom so evident in human experience. Out of this freedom the natural order produces delightful birds, such as the red cardinal that often perches outside my window on the branches of the beautiful dogwood I planted many years ago. But this freedom also gives rise to the disgusting Ichneumonidae and the naughty cat that tortures its lunch before eating it. Why is this freedom, embodied in the natural order, so much more troubling than the freedom that human beings possess—a freedom that has given rise to both hospitals and concentration camps, violins and guillotines, poetry and pornography?

THE SLIDE TO AGNOSTICISM

The Darwin described above was not a crusader against Christianity. Nor was he part of a conspiracy to destroy belief in God. He was, rather, a *reluctant* convert to evolution and ultimately agnosticism. His spiritual journey was at odds with fundamentalism, which holds that true seekers will inevitably find its version of faith. To fail to find this faith can only mean that one is not truly seeking; to *abandon* faith is simply perverted; and to create a theory that might compel people to reject faith is simply evil. In the eyes of these critics, who believe passionately that Satan is everywhere at work trying to turn people from their truth, Darwin is nothing short of an agent of the devil.[31]

Darwin eventually lost his childhood faith, but it was long after his fateful voyage aboard the *Beagle*. And although his faith in the creationist explanation for origins was undermined by his scientific work, the heart of his Christianity was destroyed by concerns much closer to home.

Darwin, like most thoughtful believers, found the Christian concept of hell—a secondary doctrine that even many conservatives reject—difficult to reconcile with the more central concept of God's love. Just as there was something theologically repugnant about God creating cats to torture mice, even briefly, there was something even more appalling about a God creating an eternal torture for those unwilling or, like Darwin, *unable* to believe. When his father died without any religious faith in 1848, Darwin confronted the reality that Christian doctrine taught that his father was now a permanent resident of hell, at the beginning of an endless torture.

Darwin became convinced that an eternal hell was more than simply a troubling and implausible concept, a cosmic parallel to the Ichneumonidae:

"I can indeed hardly see how anyone ought to wish Christianity to be true," he wrote near the end of his career, for "the plain language of the text seems to show that the men who do not believe, and this would include my Father, Brother and almost all of my friends, will be everlastingly punished. And this is a damnable doctrine."[32] Darwin's religious struggles distressed his beloved wife, Emma, as she considered the prospects of being separated eternally from her increasingly unorthodox husband. Darwin respected Emma's consistency in her faith and was troubled by the space his creeping unbelief opened between them.

In the final analysis, however, the event that did the most to destroy Darwin's faith was not his concerns about the legitimacy of hell. It was not the growing implausibility of creationism or his embrace of evolution. It was the death of an innocent and beloved child a brief three years after the death of his father.

Darwin, from birth to death, was a family man, devoted at first to his parents and siblings, then to his wife, and finally and most dramatically to his children. He had ten children, all of whom were raised in the comfortable security of Downe House, just a short distance from the family church at Downe, Kent, where he hoped to be buried. By all accounts Darwin's family life was rich. His children often accompanied him on walks, crawled onto his lap while he was working, and generally filled his home and his life with laughter. One of them, Annie, held a particularly special place in his heart.

In 1851, at the age of eleven, Annie contracted a childhood illness, possibly tuberculosis, and began what was to be a short battle for her life. The local physician dropped by several times, as did the parish priest. Emma spent much time in prayer, asking God to spare Annie's life. Charles struggled mightily. On the one hand, Annie's fight for her life was the struggle for survival that was the way of all biological life. Nobody understood that better than he. But he still believed in God, and hidden beneath the decaying vegetation of his once vibrant faith was the residue of an enduring conviction that a good and beneficent God was in control. This God cared about the fall of sparrows, the hairs on our heads, and the health of our children. "Suffer the little children to come unto me," said Jesus, when his associates would shoo them away.

Little Annie Darwin, the jewel of Charles and Emma's remarkable family, passed away on April 23, 1851. Emma memorialized Annie by creating a box of her special possessions, which she opened when the empty space created by Annie's passing seemed to grow too large. Darwin's great-grandson Randall Keynes has lovingly told this story in *Darwin, His Daughter and Human Evolution.*[33] Darwin, as befits the author of one of the world's most important books, processed his grief through writing: "We have lost the joy of the household, and the solace of our old age," he wrote on April 30, 1851.[34]

THE BODY SNATCHERS

In the final analysis, one of the greatest scientists who ever lived, the architect of the worldview that countless Christians believe was inspired by Satan to destroy their faith, the thinker who did more than anyone to drive natural theology from intellectual discourse, lost his faith when his daughter died. Darwin's belief in God weathered the theological storms brought on by the Ichneumonidae, the sadistic cats, and the webbed feet of the upland geese. He understood that those features of the natural world could be reconciled with belief in God as creator. He is followed in this belief by the majority of theologians who have reflected on these problems and concluded that evolution by natural selection is not incompatible with belief in God as creator.

But Christianity is not fundamentally about how God created the world and its many interesting creatures. Christianity is about the extraordinary claim that God loves those creatures and cares deeply about their welfare. This, alas, is undeniably difficult to square with the death of a child. Darwin's diaries, notebooks, personal correspondence, and other writings reveal the unfolding patterns of his thoughts on religion. The evidence suggests a lifetime of complex wrestling with issues of faith. His belief in God waxed and waned, but took a severe blow when Annie died.

This is not to claim that Darwin's religious faith ever completely died. He had important personal and social reasons to hang on to belief and never joined his contemporaries in their attack on the church. He continued to support his local church financially and helped with parish work, but on Sundays he went for a walk while his family was at worship. He never embraced atheism. And even within his controversial theory he continued to find room for God. In a beautiful and often quoted passage at the end of *On the Origin of Species,* Darwin wrote:

There is grandeur in this view of life, with its several powers, having been originally breathed into a few forms or into one; and that, whilst this planet has gone cycling on according to the fixed law of gravity, from so simple a beginning endless forms most beautiful and most wonderful have been, and are being, evolved.[35]

Darwin's wish was to be buried in St. Mary's churchyard at Downe next to the bodies of his children who had died. It was a place he called "the happiest on earth." But by the time of his passing in April 1882 at the age of seventy-three he had become an international symbol. Reposing in the graveyard of a humble parish would not do. Darwin's allies saw in his work the foundations of a welcome new social order, one in which science replaced religion as the dominant cultural authority and traditional social straitjackets were cast aside. This transition was effectively symbolized by a funeral that was a state occasion and a burial in Westminster Abbey, where other influential—if more orthodox—British luminaries were laid to rest. Darwin's interment in Westminster Abbey, next to the imposing statue marking the grave of the great Isaac Newton, was an emphatic statement that a new order had arrived. "Darwin's body," penned biographers Adrian Desmond and James Moore, "was enshrined to the greater glory of the new professionals who had snatched it."[36]

Society and the world, at least for those who captured Darwin's vision, had been naturalized. Throughout Britain, the power of the clergy continued a decline begun even before Darwin set foot on the *Beagle*. Eventually the church became a minority voice in an increasingly secular, pluralistic society. Throughout nature the explanatory power of theology was in similar decline, as scientific explanations displaced more traditional religious ones. Theology, however, was reeling under the impact of a far more serious crisis that had nothing to do with Darwin, brought on by radically new biblical scholarship coming out of Germany.

CHAPTER 2

A TALE OF TWO BOOKS

Darwin," writes Richard Dawkins, "made it possible to be an intellectually fulfilled atheist."[1] Such claims, by our leading public intellectual, have earned Dawkins his nickname: "Darwin's rottweiler." The label is a diplomatic downgrade from one attached to the kinder, gentler Thomas Huxley a century earlier—"Darwin's bulldog"—and derives from Dawkins's enthusiastic, in-your-face promotion of all things Darwinian.

Dawkins, the Charles Simonyi Professor for the Public Understanding of Science at Oxford University, is the world's leading popularizer of evolution. He has written many influential books, starting with the classic *The Selfish Gene* in 1976, then *The Blind Watchmaker* in 1987, and, two decades later, his 688-page opus, *The Ancestor's Tale*. One of his staunchest critics says he is "as articulate as anyone alive."[2]

Although Dawkins's writings are mainly science exposition at its best, he clearly has an antireligious ax to grind and often concludes his books by musing about how scientific accounts of origins are superior to their religious counterparts. At the end of *The Ancestor's Tale,* for example, he writes: "My objection to supernatural beliefs is precisely that they miserably fail to do justice to the sublime grandeur of the real world. They represent a narrowing-down from reality, an impoverishment of what the real world has to offer."[3]

Not surprisingly, religious believers have been at war with Dawkins for some time, a conflict escalated by his recent work *The God Delusion,* an aggressive diatribe against religion. His writings and public appearances insult Christians on two fronts—their cherished beliefs and their intelligence. Responses to Dawkins tend to be more restrained and include *Dawkins' God* and *The Dawkins Delusion* by an Oxford colleague, theologian Alister Mc-Grath,[4] and a lengthy chapter in *The Oracles of Science: Celebrity Scientists*

Versus God and Religion.[5] Dawkins's conservative critics consider him a well-defined enemy. Phillip Johnson, speaking for most of them, accuses him of being "scientifically absurd and morally naive."[6] Christians inclined to think evolution is a Satanic conspiracy see him as downright sinister. If in his next public appearance horns suddenly grew out of Dawkins's head and he announced that he was the Antichrist, come to complete the task of destroying religion begun by Charles Darwin a hundred and fifty years ago, some Christians would nod knowingly and say, "I thought so."

Dawkins and his colleagues-in-arms—Steven Pinker, Sam Harris, Peter Atkins, Francis Crick, Steven Weinberg, and Daniel Dennett—fret over the intellectual trajectory of the twenty-first century. Science has not captured the heart and mind of the culture, as they had anticipated, and religion, after a century of steady retreat, has come roaring back with a vengeance, especially in the United States. And the religion roaring back is the worst kind—Bible-reading (or at least Bible-thumping), miracle-believing, born-again, evolution-bashing Christianity. Science finds itself in an uncomfortable and unfamiliar defensive role, reduced to defending hard-won territory against the philistines.

Dawkins, Dennett, and company, who call themselves "brights"[7] to distinguish themselves from "dims," who believe in God, are the contemporary champions of the secularist worldview that captured France in the eighteenth century, invaded England in the nineteenth century, and frightened America in the twentieth. But Dawkins's use of evolution to undermine religion differs from what happened in nineteenth-century England, when Darwin's new theory was first introduced.

GOD'S FUNERAL

In the first place, the nineteenth-century secularism that Dawkins celebrates was not driven primarily by science, but by forces *internal* to religion, especially German biblical scholarship. Science played but a small role, and an ambiguous one at that. Ironically, it was those most familiar with the Christian Scriptures and the history of the early church who initiated and encouraged the move toward secularism. David Friedrich Strauss, for example, produced a critical and scholarly analysis of the Bible titled *The Life of Jesus Critically Examined.* This book raised thundering questions about the reliability of the Bible.[8] Appearing in 1835 in German, while Darwin was measuring finch beaks on the Galapagos Islands, Strauss's monumental work was translated into English by the novelist George Eliot in 1846. Over the next decades the book exerted an unprecedented influence on the study of Jesus's life. Strauss sought to discover the "historical Jesus" using both the gospels and extrabiblical sources. In doing so, he undermined the validity and historical reliability of the gospels, spreading crises of faith across Europe like a plague. In contrast, many of those same readers, as well as scientists and even clergy, were reading *On the Origin of Species* without getting the least bit sick.

In the second place, the nineteenth-century loss of faith was not received as a liberation—an "intellectual fulfillment," to paraphrase Dawkins. The soldiers of doubt that came blasting through the walls of England's many houses of worship, from Westminster Abbey to the humble parish church where the Darwins worshiped, were enemies, not liberators. Most nineteenth-century Christians who lost their faith were deeply troubled by the experience. Some were plagued by apocalyptic visions of a post-Christian Europe. Like Darwin, who fantasized about the discovery of documents corroborating the New Testament stories and chasing away the demons of his doubt, nineteenth-century unbelievers did not enjoy the disintegration of their faith.

The most eloquent of these laments is Matthew Arnold's "Dover Beach," written in 1867. Arnold concludes his poignant masterpiece by comparing the European loss of faith to a tide going out:

> The Sea of Faith
> Was once, too, at the full, and round earth's shore
> Lay like the folds of a bright girdle furled.
> But now I only hear
> Its melancholy, long, withdrawing roar,
> Retreating, to the breath
> Of the night-wind, down the vast edges drear
> And naked shingles of the world.
> Ah, love, let us be true
> To one another! for the world, which seems
> To lie before us like a land of dreams,
> So various, so beautiful, so new,
> Hath really neither joy, nor love, nor light,

Nor certitude, nor peace, nor help for pain;
And we are here as on a darkling plain
Swept with confused alarms of struggle and flight,
Where ignorant armies clash by night.[9]

The novelist and poet Thomas Hardy penned "God's Funeral" around 1909, even as he was being engulfed by the unwelcome fog of atheism. Like Arnold, Hardy captures the sense of loss and hopelessness brought on by the emerging crisis of faith:

So, toward our myth's oblivion,
Darkling, and languid-lipped, we creep and grope
Sadlier than those who wept in Babylon,
Whose Zion was still abiding hope.[10]

Hardy, Arnold, and their fellow Victorians who attended God's funeral found no intellectual fulfillment in the ideas that made belief in God optional, redundant, or even unacceptable.[11] It would be decades before people like Dawkins would upend the Victorian sentiments and try to spin the nineteenth-century loss of faith into something wonderful and liberating. Dawkins's lament, of course, is that God, like Jesus in the New Testament, didn't stay dead.

TWO BOOKS

These European intellectual currents traced different courses as they made their way to America. It is instructive to compare the reception of the two great books from Europe, Strauss's *The Life of Jesus Critically Examined* and Darwin's *On the Origin of Species*. Both were destined to exert great influence on Christianity, although in dramatically different ways and on different schedules.

Strauss's work, a part of the movement already under way known as *higher criticism,* generated enormous controversy. Immediately rejected by conservative Christians, it spawned a backlash that split Christianity into two camps—liberals, who accepted it, and fundamentalists, who did not. *On the Origin of Species* produced a more complex and organic reaction. Destined to eventually be at the heart of a national crisis in the public schools, evolution was initially dismissed by many American religious leaders as scientifically absurd and unlikely to endure. Within a decade, however, the scientific community had embraced evolution, muting claims it was absurd and motivating thoughtful Christians, including many conservatives, to make peace with Darwin's new theory. Many concluded that evolution offered no clear threat to faith. Flexibility in interpreting both the theory and the Bible enabled the reduction and even elimination of apparent contradictions.

No such peace was to be found with higher criticism, which appeared to be making a full frontal assault on the reliability of the Bible. Strauss and his colleagues brought an unprecedented historical and literary approach to the Bible, treating it as any other ancient document rather than the sacrosanct "Word of God." The results were disturbing. Serious questions were raised about everything from miracles to the very existence of Jesus.

The gospels, noted the critics, disagree on such basic history as Jesus's resurrection. Matthew places two women at Jesus's tomb, Mark places three, Luke more than three, and John only one. What is going on here? Now that we understand the importance of history, how can readers put faith in the historicity of an event chronicled by such unreliable reporters? And what was the big deal about Christian miracles when miracle stories were so common outside of Christianity? Pythagoras, for example, was said to be the son of Apollo, born of a virgin, and to have calmed storms and visited the dead in Hades.[12] Why do we privilege such claims when we find them inside the Bible and reject them when we find them outside of it?

Strauss's bombshell, despite the author's assurances that his work was in the service of Christ, riled his colleagues in Germany and got him fired from Tübingen University. As his inflammatory text made its way across the Atlantic Ocean to America, religious militias lined up along the coast from Maine to Florida trying to prevent it from coming ashore and taking up residence within evangelicalism. In contrast, Darwin's *On the Origin of Species* disembarked with less fanfare and soon found some evangelical doors open to it—doors that had been slammed in the face of higher criticism.

DARWIN COMES TO AMERICA

On the Origin of Species arrived in America in 1860. Considered an important new scientific work, it was reviewed by leading scientists in influential opinion journals like the *Atlantic Monthly* and the *American Journal of*

Science and Arts. America's leading biologist, Louis Agassiz, of Harvard, described Darwin's theory as "a scientific mistake, untrue in facts, unscientific in its methods, and mischievous in its tendency."[13] Critics like Agassiz empowered Christians, at least initially, to reject Darwin's theory on scientific grounds.

Darwin's book sold well. It slowly began to win the loyalty of biologists and reshape the life sciences, but it did so without apparently disrupting the prior religious commitments of those who embraced it.[14] It would be a half century before William Jennings Bryan and Clarence Darrow would spar about evolution at the Scopes trial, and a full century before America's fundamentalists would be united en masse against it, certain it had been conjured in hell by Satan himself.

The steadily evolving complexity of America's response to Darwin resulted from multiple ambiguities in play.[15] For starters, there was no consensus on exactly how Darwin's theory should be understood. Evolution in several forms was "in the air," and Darwin's contribution was in some ways just a well-documented presentation of ideas that had been bandied about for decades. Darwin's own theory, with the unique role assigned to natural selection, had been circulating quietly for two decades and had even been independently proposed by another naturalist named Alfred Wallace. Most biologists were soon convinced that evolution had occurred, more or less as Darwin described in *On the Origin of Species,* but they were skeptical that the process of natural selection, all by itself, was up to the task of turning an amoeba into a proper Victorian.

Complementary ambiguities attended the interpretations of Genesis and whether evolution was necessarily incompatible with creation. Between multiple explanations for how evolution worked—some of which were congenial to Christianity—and various interpretive schemes for Genesis, there was simply no need for Christians to get alarmed about Darwin's American debut. Earlier developments, in fact, had even prepared the way for evolution through a series of compromises on things like the age of the earth or the extent of Noah's flood. Such compromises had opened space for new ideas. Controversy was also muted by the fact that the great scientific authorities of the day were mostly all Christians and not inclined to put any antireligious spin on new scientific developments.

Prior to Darwin, the influential Swedish farm boy turned botanist Carolus Linnaeus (1707–78), who gave us such delightful labels as *Homo trog-*

lodyte (cave man), had subscribed to a clearly religious concept of origins: God created two of each species, which then dispersed to populate the globe. Species could neither evolve nor go extinct, so Linnaeus's famous labeling exercise simply cataloged what God had done a few thousand years ago. Biblically influenced views like those of Linnaeus were modified under the pressure of accumulating evidence. Charles Lyell (1797–1875), whose *Principles of Geology* had shaped Darwin's views while aboard the *Beagle,* believed the scientific evidence indicated that God had created species at multiple locations and on numerous occasions, a strongly creationist but decidedly unbiblical explanation.[16] Louis Agassiz (1807–73), a world-class ichthyologist and one of America's first great scientists, believed that God had created species in large numbers, repopulating the earth after various divine tantrums like the great flood of Noah. These and other views with meaningful connections to the Christian understanding of creation were endorsed by leading naturalists during a time when biology was still developing. The variety of such theories made it impossible to assess the degree to which new science challenged religious understandings of origins. Furthermore, Darwin did not yet tower over others of the nineteenth century as the key scientist, so his authority was not considerably greater than that of scientists promoting other views.

Confronted with multiple theories about geology, biology, and Genesis and their relevance to each other, scientists and clergy alike were liberated to think creatively about origins. Growing evidence that the earth was ancient pushed the origin of the earth back in time; the discovery of fossils belonging to creatures that went extinct long before humans appeared forced a reinterpretation of the chronology in Genesis. And although such developments put Linnaeus's creationism to rest, there were other options available at the time.

For those who would defend the Genesis creation story as more than a myth, there were two interpretations of it that preserved at least a mutant form of biblical inerrancy: the *day-age* theory and the *gap* theory. Both achieved some currency during the nineteenth century in response to the growing geological evidence for the great age of the earth.

The day-age theory accommodated the great age of the earth by converting the days of creation in Genesis into geological epochs. There was biblical license for making this move. The Hebrew word for "day," *yom,* sometimes referred to a period of time rather than an interval of twenty-four hours. In

Psalm 90:10, which contains the word *yom,* we read: "The days of our life are seventy years" (NRSV). Modern English expressions like "in this day and age" and "your day will come" reflect similar usage. William Jennings Bryan admitted on the Scopes trial witness stand that he subscribed to the day-age theory, to the delight of Clarence Darrow and the chagrin of the more literalist members of his fan club.

A version of the day-age theory appeared in 1778 when a leading French intellectual bearing the ponderous name Georges-Louis Leclerc de Buffon published *Epochs of Nature.* Buffon was one of the first to surmise that the earth had a long, complex evolutionary history and had not been created a few thousand years ago looking much as it does today. He proposed that the earth originated when a comet collided with the sun and ejected material out of which the earth formed. This molten material needed more than a few thousand years to cool to its present temperature, perhaps as much as three million years.[17]

Criticism rained down on Buffon as the guardians of biblical orthodoxy and even Voltaire reacted to this new theory of earth history. To pacify the religious critics Buffon divided the newly extended history of the earth into seven epochs, sequenced in a way that lined up with his proposed evolution of the planet. The seven epochs may have been a charade, but it provided a scheme by which a creative interpretation of Genesis could be reconciled with what was to be a steadily increasing age for the earth.

In the early nineteenth century the day-age scheme became enormously popular through the writings of Hugh Miller, a respected religious leader gifted with "elegance, grace, and wit."[18] This interpretive scheme found broad application as religious believers, including geologists, sought to preserve the historicity of the Genesis account of creation, even as their understanding of that creation underwent cataclysmic change. The day-age theory lives today in the work of several leading fundamentalists, most notably Hugh Ross, who heads up the Christian apologetics ministry Reasons to Believe. Ross, who has a Ph.D. in astronomy, promotes an integration of natural history and the Genesis creation story utilizing the day-age concept to reconcile his literal reading of Genesis with evidence for the great age of the universe. Other creationists view Ross with suspicion, however, and lament his capitulation to a flawed scientific perspective and compromised reading of the Bible.

The second strategy for dealing with the age of the earth was the gap theory, so-called for its insertion of a great historical gap between the first and second verses of Genesis. In the first verse in Genesis we read: "In the beginning God created the heavens and the earth." In the second verse we read: "And the earth was without form and void." The gap theory interprets the first verse in Genesis as referring to a *prior* creation event. The second verse refers to the most recent creation. We thus have an undefined epoch between the two into which almost anything can be inserted. If geologists need a few billion years of history before humans appear, we can insert that history neatly between verses 1 and 2.

Two theologically trained geologists popularized the gap theory at the beginning of the nineteenth century, Thomas Chalmers in Scotland and William Buckland in England. Motivated by geological developments and rationalized with the same sort of scriptural vagaries exploited by the day-age theory, the gap theory provided space for a different creation before the present one. Chalmers and Buckland thus developed a second biblically acceptable way to deal with the emerging geological evidence for an ancient earth. Most practicing geologists at the time took the Bible seriously. But, since they also took geology seriously, they were forced to find space in the Bible's hermeneutical holes for the latest discoveries about the earth.

Textual license exists for the gap theory. The Hebrew grammatical construction does not require that God is creating "out of nothing," but rather allows the translation that God is working with preexisting materials. In the more recent and literal translation of the New Revised Standard Version, Genesis 1:1–2 reads: "In the beginning when God created the heavens and the earth, the earth was a formless void and darkness covered the face of the deep." The most straightforward interpretation of these verses is that there was something in place on which God was working at the time the story begins. We are not told what it was or how it came to be a "formless void," but there is no obvious reason that it could not be the residue of a previous catastrophe along the lines of Noah's flood or the destruction of Sodom and Gomorrah.

The gap theory became very popular and eventually made its way into the influential *Scofield Reference Bible,* first published in 1909, which has since sold over two million copies. The Scofield Bible, still available from Oxford University Press in a revised edition, contains copious study aids prepared by a biblical scholar named Cyrus I. Scofield. This study Bible was the definitive Scripture for many fundamentalists throughout the twentieth century. Oddball interpretations of various biblical passages showed up in the study aids and acquired an almost canonical status by virtue of being included in the volume. One of the more alarming examples of this involves the episode in Genesis that follows the great flood. Noah's son Ham is cursed for improprieties with his naked, drunken father. The curse includes a reference to slavery, and Scofield's notes in the earlier versions suggest that these verses offer a justification for the abuse of black people. The curse on Ham's descendants was supposedly dark skin and a secondary role as servants of white people, an interpretation used to rationalize slavery.

In the notes accompanying the first chapter of Genesis, Scofield references multiple biblical passages that "clearly indicate that the earth had undergone a cataclysmic change as the result of divine judgment."[19] He goes on to say, "The face of the earth bears everywhere the marks of such a catastrophe," and suggests that the catastrophe resulted from "a previous testing and fall of angels." I can remember reading these notes in my father's Bible as a child, impressed that Scofield knew so much about how God had done things and wondering what marvelous events must have attended the testing and fall of angels.

The Scofield Bible also reproduced Bishop James Ussher's seventeenth-century biblical chronology, which stated that the creation week described in Genesis occurred in the year 4004 BCE, but after the gap inserted between the first two verses to accommodate a prior creation. In the early versions of the Scofield Bible that date appears in a column in the center of the first page of Genesis.

The day-age and the gap theory are tools to reconcile the great age of the earth with a literal reading of Genesis, and millions of Christians found them entirely adequate. However, there were reasons why this interpretive strategy might not even be necessary. Multiple elements in the Genesis stories of creation suggest a figurative or symbolic, rather than a literal, reading. The angel with flaming sword guarding Eden's gate, for example, struck many as a mythological element, especially as it implies that the Garden of Eden is still present somewhere on earth. The talking serpent, God strolling through the garden in the evening with Adam, and the rib surgery to make woman all strained the plausibility of a purely literal reading. Even some literalists will concede that these elements are laden with symbolism and allegory.

The history of interpreting Genesis also reveals a diversity of readings, even before there were pressures from science. More than a millennium before Darwin, to take one example, St. Augustine took time off from obsessing about the evils of sex to write a commentary suggesting that the Genesis creation week was not to be taken literally.[20] Augustine saw no reason to suppose that God would organize his work week as humans do and then, when he was done, take a day off to do God knows what.

All these factors were in play in late nineteenth-century America when Darwin arrived. There were several entirely legitimate readings of Genesis available, from Augustine's allegorical approach, to Buffon's geological interpretation, to the traditional six-day creationism. Some of these readings were compatible with evolution, and some were not. Without some galvanizing event or charismatic leader to rally Christians to the cause, there was simply no need for them to take up arms to defend the integrity of the Bible against any imagined assault from evolution.

In addition to the lack of unanimity on how to read the biblical creation stories, America's response to *On the Origin of Species* was further shaped by the ambivalence of biologists toward Darwin's explanation for evolution. Darwin, most agreed by 1870, had amassed compelling evidence that evolution had occurred. That all life was the result of constant change over time from common ancestors became known as the "fact" of evolution. But Darwin's theory purporting to explain *how* this occurred was another matter. There was simply no consensus that natural selection was the mechanism of evolutionary change. Many, in fact, were skeptical that an undirected chance process could account for the astonishing diversity and creativity of the natural world.

Some who accepted evolution suspected that unknown mechanisms were at work, perhaps guided by God. In any event, the historical fact that evolution had occurred was easily separated from any particular theory of how it had occurred. We must keep in mind that the full name of Darwin's book was *On the Origin of Species by Means of Natural Selection, Or the Preservation of Favoured Races in the Struggle for Life*. Darwin's great work must thus be considered from two separate and separable perspectives: common ancestry as an empirical fact, and natural selection as a theoretical explanation for that fact.

Biologists today consider the common ancestry of all life a fact on par with the sphericity of the earth or its motion around the sun. They note the mountain of evidence that all life came from a common ancestor several billion years ago. Evidence from comparative DNA, the fossil record, the geographical distribution of life, comparative anatomy, and other data point to a common ancestor. The evidence is so compelling that even some dedicated anti-evolutionary intelligent design advocates have grudgingly conceded on this point, even as they reject just about every other aspect of evolution. Michael Behe, for example, author of the important and readable intelligent design classic *Darwin's Black Box,* suggests that all the life-forms on the earth may have come from a single common ancestor—an "über-cell" with all the "designed systems" needed to give rise to the entire panorama of all the life that has existed.[21] His fellow intelligent design theorist William Dembski has speculated that maybe God "front-loaded" everything into the big bang.[22] Belief in a common ancestry for all life, from Darwin's day to our own, has never entailed accepting any particular mechanism for how that common ancestor managed to give rise to the great panorama of life that has graced our planet.

Without the insights of genetics, which were not incorporated into Darwin's theory until well into the twentieth century, there was no knockdown argument for natural selection as *the* mechanism for evolution. Certainly his fellow biologists were not knocked down by natural selection as the mechanism to get a Victorian from an amoeba. And although they were convinced that this was indeed how Victorians had arisen, they were not sure that something so feeble and obviously purposeless as blind natural selection could accomplish that remarkable task.

ALTERNATIVE EXPLANATIONS FOR EVOLUTION

Two alternatives challenged natural selection as the driving force of evolution, and both of them nestled comfortably into the worldview of the nineteenth century, with its orientation toward progress and purpose. The first had a pedigree going back to the French naturalist Jean-Baptiste Lamarck (1744–1829) and was based on the inheritance of acquired characteristics. Lamarck was an evolutionary voice in the wilderness, preparing the way for Darwin. His central idea of evolution, however, despite eventually being proven completely wrong, resonated so well with nineteenth-century intuitions that it would be a century before biologists had fully expunged its heresy.

Lamarck noted the obvious fact that organisms develop adaptations that help them function. A blacksmith, for example, develops large muscles, which aid him in his work. A concert pianist develops nimble fingers. Giraffes stretch their necks to reach food, and dancers become more graceful as they perform continually on stage. Lamarck proposed that such traits, developed to better negotiate the challenges of life, could be passed on to offspring. If a mother giraffe had been especially industrious in stretching her neck to reach food at inconvenient heights, her offspring would be born with the potential for longer necks than their peers. The son of an industrious blacksmith would be destined to develop a robust physique; the daughter of a nimble ballerina would be born with enhanced grace and agility; a politician who learned how to fool people would have similarly talented children. In this way, lineages could experience genuine progress.

Those of Darwin's generation believed in nothing so much as progress, enamored as they were with the elevated stature of their own culture. Lamarck's theory reinforced Victorians' belief that their achievements—in industry, literature, music, and so on—were the result of diligence and hard work, as each generation benefited from the efforts of the previous one. As an important corollary, such a belief also relieved concerns about caring for the poor, whose unfortunate circumstances could be rationalized as the result of laziness. Generations of lazy slackers had been passing down a deteriorating work ethic for centuries, and now the situation was beyond repair.

Lamarck's theory invested natural history with a moral dimension. Progress was good, the reward for diligence and hard work, whether it be a finch pecking with greater vigor on the Galapagos Islands or a human ancestor showing courage and vigor in the face of life's challenges. The common-sense character of Lamarck's explanation, especially in the decades before the genetic basis for inheritance was understood, was deeply intuitive, and most naturalists, including Darwin, accepted it. Even today we must admit that it is far more appealing than the blind and purposeless selection processes that eventually came to define orthodox Darwinism. Unfortunately, it isn't true.

Another alternative to natural selection available at the end of the nineteenth century was *orthogenesis,* which was about as anti-Darwinian an evolutionary explanation as one can imagine. The term means "evolution in a straight line" and refers to the idea that species evolve along a path specified by something akin to a blueprint. Evolution by these lights has "momentum" and moves species forward according to a plan of some sort, which many were quite happy to ascribe to God, now residing comfortably in deist quarters. There were variations on this basic theme, but the central notion was that evolution is driven by a force independent of the environment or any other conditions associated with species.

Orthogenesis was partially inspired by interesting similarities between evolution and human development, a topic explored at great length by many nineteenth-century thinkers. Human beings, for example, begin as simple one-celled organisms—the fertilized egg—and develop steadily in complexity, first in the womb and then outside of it. They reach adulthood, enter a period of stability, then begin to deteriorate, and finally die. Likewise, species originated with one-celled common ancestors, increased steadily in complexity until they reached a period of stasis, and finally went extinct. The similarities were provocative, to say the least.

Human development was deeply mysterious and poorly understood at the time of Darwin, and is still deeply mysterious in many ways. But there was no denying that it occurs regularly and reliably, driven by mechanisms completely unknown to Darwin's generation and not fully known to ours. The champions of orthogenesis simply and reasonably invoked an analogous mechanism as the driver of evolutionary change. There was ample evidence that could be offered in support.

Consider the many things that have evolved that don't seem remotely adaptive. Some, like the interesting patterns on butterfly wings, seemed pointless. Others, like the gigantic antlers on the Irish elk, were maladaptive and thought to have contributed to the extinction of that species. Such evidence, if legitimate, was difficult to square with evolution by natural selection, which would hardly endow elks with antlers that would hasten their demise. On the other hand, if there was such a thing as "evolutionary momentum," it was easy to see how a developmental process that produced antlers in the first place could "overshoot" and leave the unfortunate elk with an unmanageably large rack. A human parallel might be a person increasing in weight in the movement toward adulthood, but then moving inexorably into a state of great obesity, leading to health problems and ultimately to death.

Orthogenesis was, and is, intuitive. It is such a natural misunderstanding that, when I teach evolution, I make a point of emphasizing that "evolution is not a force like gravity, constantly prodding everything to evolve." Nevertheless, it is inevitable that at least one skeptical student will raise a hand and challenge me by asking: "If we evolved from apes, then why are there still apes?" Or, "If evolution is true, then why are we not still evolving?" Both these questions assume that evolution is some kind of mysterious force that, like gravity, propels species along some evolutionary pathway. Orthogenesis postulated just such a force moving the evolutionary process along, much like the force that drives human development from conception to adulthood.

Both Lamarckism and orthogenesis offered alternatives to natural selection, especially for audiences obsessed with progress. Both theories were compatible with religion, viewed from the right angle. The blueprints guiding orthogenesis *could* be viewed as the handiwork of God. The progress of organisms under Lamarckism *could* be viewed as moral imperatives, with every creature investing in its own creative and purposeful advance, for the good of its offspring.

Historian of evolution Peter Bowler has examined the non-Darwinian options available around 1900 in his book *The Eclipse of Darwinism*. He concludes that the popularity of these alternatives that were eclipsing natural selection "all originated in a long-standing tradition that organic development must be an orderly process controlled by laws inherent in life itself." Identifying and understanding these laws had the potential to transform biology into the same sort of rigorous, mathematical science as physics, which was taken as the ideal. What we now call Darwinism, in contrast, with its emphasis on boring, blind, and lifeless natural selection, seemed "moribund and incapable of furthering biological research."[23]

BACK TO THE TALE OF TWO BOOKS

I have sailed briefly into these non-Darwinian waters to make the point that the Darwin who arrived in America was not the same fellow who had written *On the Origin of Species*. By 1875 evolution as a historical fact had been established to the satisfaction of most scientists as well as educated people who had taken the time to absorb Darwin's argument. The vastness of natural history, both geological and biological, no longer threatened people as it once had. But nobody had a clear idea of how evolution occurred, largely because so little was known of genetics. In the absence of a solid, empirically grounded theory of evolutionary change, speculative hypotheses found a ready audience. And some of these hypotheses were quite congenial to both religion and common sense. No one could argue convincingly yet that the central character in the evolutionary story was that blind and indifferent pruner called natural selection.

Ambiguities about evolution coexisted with ambiguities about biblical interpretation. For evolution to conflict with the Bible, these ambiguities would have to resolve in a specific way that was genuinely incompatible. We can certainly select a biblical interpretation that will conflict with a particular explanation for evolution. But why would we want to do that? Absent a revelation from God commanding such a cantankerous move, there is simply no reason to do this. Blessed are the peacemakers, said Jesus, not those who go around manufacturing controversy.

THE TINY SEED OF CONTROVERSY

Unfortunately, there was one American religious leader who did get a command from God to make just such a fuss about evolution. Her name was Ellen White, and the small anti-evolutionary flame she kindled over a century ago has all but engulfed evangelical science in America.

Ellen White (1827–1915) and her family were part of a cult that followed an apocalyptic preacher named William Miller, who predicted that Jesus would return on October 22, 1844. Needless to say, Jesus did not return as predicted, and the mass gathering of frustrated faithful dispersed. Many returned to their former, more traditional denominations. Shortly after this "Great Disappointment," as it became known, White began to experience vivid religious visions. Many believed that God was speaking to her, and she soon emerged as an important religious leader in a new sect known as the Seventh-day Adventists. In 1863 the Adventist religious group was formally established, with White as one of the founders. Her followers consider her writings to be inspired and treat them with great respect, almost on par with the Bible. One of White's first visions was of the Seventh-day Adventists marching into heaven, unaccompanied by the apostate Christian groups.

In 1864, five years after the publication of *On the Origin of Species,* White wrote that God had given her a vision of the actual creation: "I was then carried back to the creation and was shown that the first week, in which God performed the work of creation in six days and rested on the seventh day, was just like every other week."[24] These and other prophetic writings by White rooted the Adventist movement firmly in the soil of young-earth creationism.

White's influence on American culture was limited, however, by the small size of the Adventist sect, which numbered just 140,000 members early in the twentieth century. (In contrast, there are now 14 million Adventists in 202 countries.) Most Christians view Seventh-day Adventists with suspicion, put off by their apocalypticism, odd dietary laws, and theology of the Sabbath, according to which they worship on Saturday, when they should be mowing their lawns. And many Christians have long considered White somewhere between a false prophet and a mentally deranged person, or perhaps even a mentally deranged false prophet.

Despite Adventism's cultural insignificance in nineteenth-century America, the modern creationist movement was gestating within its eccentric theological womb. By the early twentieth century a self-taught Adventist geologist named George McCready Price would recast White's vision of Noah's flood in scientific terms. The achievement inspired John Whitcomb and Henry Morris to write *The Genesis Flood,* and the rest, as they say, is history.

But flood geology was irrelevant when Darwin arrived in America in the middle of the nineteenth century, and there was limited active opposition. Scientists had various interpretations of evolution, some of which were theologically benign. Likewise, theologians and biblical scholars were not united behind a reading of Genesis incompatible with evolution. With the exception of the marginalized Adventists and a few Protestant conservatives, there was far less fuss than might have been expected. A mass movement opposing evolution was a half century away.

THE BIRTH OF FUNDAMENTALISM

There was, however, a mass movement opposing the other book from Europe, Strauss's *Life of Jesus Critically Examined.* Strauss's controversial book had conservatives wringing their hands, lamenting from pulpits, and feverishly writing refutations.

Segments of American Christianity under the influence of higher critics like Strauss abandoned the Bible as the ultimate authority for faith. Theologians calling themselves "Christian" rejected the New Testament miracles and the divinity of Christ. They treated the Bible as a purely human book. Lyman Abbott (1835–1922), for example, a prolific author and theologian who was for several years a minister in the Congregational Church, in 1892 published *The Evolution of Christianity,* offering an "updated" religion with no heaven, hell, original sin, or divine Christ.[25] Such reformulations of Christianity abandoned so many traditional beliefs that many feared there was no baby left in the tiny puddle of remaining bathwater.

To meet the rising tide of modernism, as it was known, an influential project was launched in 1909 to identify the essential core ideas of Christianity—the fundamentals—and rally Christians to protect those beliefs and keep them from being swept away by the rising tide of modernism. A lively conversation ensued. Which ideas were fundamental to Christianity and which were secondary or even peripheral? What were the issues on which Christians could disagree? Do Christians have to believe that a whale swallowed Jonah? That Job was a real character? That Noah's flood was global? That Jesus was born of a virgin? That God is a trinity? Selection of the contributors, which included many leading Protestant thinkers,[26] required identification of scholars believed to represent the best Christian thinking. The entire fundamentals project entailed engagement with traditional Christianity at all levels. The result was a four-volume set of essays titled *The Fundamentals.*

The primary target of *The Fundamentals* was obvious. Of about ninety articles in the series, fully one-third defended the Bible against Strauss and the higher critics. The rest presented doctrines, laid out apologetic arguments, criticized various "isms," and discussed world evangelism and other practical matters. Some of the essays were personal testimonies written by exemplary Christians.

Evolution in some guise appeared in about 20 percent of the essays. What was remarkable about these discussions of evolution, however, was the almost total absence of the six-day creationist viewpoint. Leading "fundamentalist" thinkers spoke approvingly of progressive creationism, historical linkages between species, and an ancient earth. There were critical comments as well, of course. One author maligned evolution by connecting it to higher criticism and called it an enemy of the Christian faith. More typical, however, were the views expressed by George Frederick Wright of Oberlin College, who claimed that the challenges from philosophy were far more serious than those from science. "Hume," he wrote, "is more dangerous than Darwin."[27]

Clearly, even leaders concerned with defining and protecting the *fundamentals* of Christianity shared no consensus on what Christians should think about evolution. This ambivalence in *The Fundamentals* offers a key insight into the history of this controversy. The fundamentalist movement, today unanimously opposed to evolution, takes its very name from this project. And yet this original generation of authentic fundamentalists was relatively unconcerned about evolution. Modern creationists should reflect on the fact that *The Fundamentals* contains no call to take up arms against evolution.

The Fundamentals succeeded in rescuing Christianity from modernism, largely because two wealthy Christian oilmen donated a small fortune to the project. They underwrote the production of the original twelve pamphlets of essays and then paid to ship almost four million copies free of charge to Christian leaders around the world. A distinct branch of Christianity known as *fundamentalism* resulted.

Much of the opposition to evolution in the early twentieth century came from the marginal and largely irrelevant Adventists, who were not invited to contribute to *The Fundamentals*. Eventually, however, the anti-evolutionary views of the Adventists migrated beyond the borders of their small sect and influenced the larger fundamentalist community, evolving into the movement we now know as *scientific creationism.*

Profound concerns about evolution emerged from a very different source, however. Darwin, it seemed, was gathering an unsavory collection of traveling companions. His central idea that nature improved species by "selecting" the more fit attracted the attention of some shady characters with rather different ideas about exactly what "fit" should mean. Aggressive militarists, particularly in Germany, invoked Darwin to justify assaults on weaker nations. Social planners claimed that programs that forcibly sterilized the "unfit" were simply good science. Empire builders rationalized the extermination of "less advanced" races as a way to improve the human species.

Eventually Darwin's name was on the lips of the architects of Nazism as they rationalized their implementation of the "final solution." As thoughtful Christians observed Darwin's shady and immoral fraternizing, it became increasingly natural for them to recoil from evolution altogether. Many, like William Jennings Bryan, were alarmed to see evolution invoked to justify the German militarism that led to World War I. Closer to home, many Christians wondered if evolution really justified a 1927 Virginia court order to sterilize Carrie Buck against her will for being "feebleminded." The "science" of evolution dropped off the radar as these social agendas loomed ever larger.

Defenders of evolution as a reliable theory of origins worked steadily throughout the twentieth century to detach Darwin's theory from its fraternity of dark companions. They had limited success. Like unwanted ghosts, the dark companions continued to haunt the theory of evolution even as it became the central organizing principle of the entire field of biology.

CHAPTER 3

DARWIN'S DARK COMPANIONS

A crime was committed while I was writing this chapter. My daughter, on a travel course to Rome, had her iPod stolen. This, of course, constituted an emergency, for iPods are as important to American teenagers as kidneys and lungs. The tiny music player had to be replaced immediately.

Apple's iPod and its associated iTunes music store have been wildly successful international consumer products, changing the way music is marketed, redefining "cool" (for the moment, at least), and raking in revenues for Apple shareholders. Apple's business model in this market has been aggressive, designed to destroy competitors. Both the iPod and the iTunes store are best-of-breed products. But neither is especially unique. Less sexy but equally effective models are available, at least for now, from Microsoft, Sansa, Sony, and a host of other market competitors. Online stores can beat the iTune prices as well, with everyone from Microsoft to Wal-Mart selling music online to a growing market of listeners.

Apple, however, knows its iPod is so totally cool that every teenager has to have one, even though there are other similar music players. Apple also knows its iTunes store is by far the most popular place to buy music, even though it is not all that different from the stores run by the competition. But to weaken Microsoft, Sony, and all the other companies who make portable music players, Apple designed iTunes so that they would play only on iPods. If you want to shop at the popular iTunes store, you have to play your music on an iPod.

Apple competes in the tough world of modern capitalism, where, at least in theory, companies making the best products for the lowest price defeat competitors in the quest for consumer dollars. Apple's goal is to kill off the competition in a sort of economic genocide so their "superior race" of products will have the market to itself. And we all benefit from this practice, at least in theory.

The capitalism practiced by Apple is much older than the theory of evolution, but as soon as Darwin's theory came on the scene, with its "selection" and "survival of the fittest" themes, obvious comparisons were invoked. The "free market" was to capitalism what "nature" was to evolution, a competitive environment rewarding excellence and weeding out inferior products. The best music player by these lights—the iPod, at the moment— is the "fittest" and wins the battle, if not the war. The less fit challengers, the Zune and the Walkman, will go extinct if they cannot evolve into something more "fit."

The upside of this competition is better products at lower prices, if everyone plays by the rules. The downside is the trampling of companies— and employees—creating products that don't survive. But because there are "rules" to the game, critics are charging Apple with a crime in response to the company's aggressive business practices. This is a peculiar state of affairs. Apple makes iPods and runs the iTunes online store. Nobody is forced to shop at iTunes—there are other places to purchase music—but if customers want to shop at the popular iTunes store, they have to buy an iPod music player. Critics accuse Apple of breaking the law, however, and argue they should be penalized, just as if they were robbing banks or dumping hazardous waste into a river.

Apple's business practices raise the fascinating question of so-called *social Darwinism,* the application of evolutionary principles to social behaviors. Conventional *biological Darwinism* provides an acceptable explanation for the origin of species: complex "fit" species evolve, survive, and prosper, while less fit competitors stagnate, die, and go extinct. Biological evolution, in its pure form at least, is purely *descriptive.* It tells us, as best it can, what happened, like a video of an event. It does not pass judgment on whether the history it describes was good or bad, just as a video passes no judgment on the event it captures.

Social Darwinism, in contrast, often has a strongly *prescriptive* component, since it applies to human behavior. Moral judgment is passed on behaviors based on how they fit into the overall Darwinian scheme. Apple—or any other corporation—is allowed to destroy the products and companies competing with it, provided it plays by the rules. If people lose their jobs and become homeless, that is acceptable, since the *process* is valued and protected for the *products* it produces. As Andrew Carnegie wrote in 1889, "While the law may be sometimes hard for the individual, it is best for the race, because it insures the survival of the fittest in every department."[1] Carnegie's key phrase, "survival of the fittest," almost universally ascribed to Darwin, actually originated with the influential British philosopher Herbert Spencer. Spencer believed that everything, from the cosmos, to society, to Carnegie's free market where Apple competes with Microsoft, steadily evolves toward some sort of perfection through a process similar to what Lamarck had proposed. Spencer's ideas were in circulation before Darwin published *On the Origin of Species* and are credited with popularizing social Darwinism, although questions exist about exactly how "Darwinian" his ideas actually were.[2]

Social Darwinism remains a controversial topic around which countless questions continue to revolve. What, exactly, does the term mean? What did Darwin think about this supposed extension of his ideas? What is the actual connection between biological evolution and social Darwinism? Do the moral prescriptions of social Darwinism really find support in Darwin's theory? To what degree was biological Darwinism invoked for propaganda purposes to buttress ideas with no connection to evolution? Was there, for example, an *actual* connection between evolution and Nazism, as a recent scholar has argued?[3] Or is this just a propaganda move to make evolution smell bad?

These questions will no doubt occupy scholars for years to come and may never be resolved. Certainly I am not going to resolve them in this brief chapter. But their resolution is not important for my purposes. I simply want to argue that the mere existence of the *concept* of social Darwinism has enormous significance for understanding reactions to evolution. That a connection can be and has been drawn between evolution and Nazism creates a disastrous public relations problem for Darwin. Such connections only further prejudice the millions predisposed to be skeptical about evolution against the theory and play into the hands of already powerful anti-evolutionary pundits.

These and the other controversies that swirl around evolution derive from the theory's great subtlety, ambiguity, and widespread applicability. We encounter evolutionary phenomena at so many levels and apply the term in so many contexts that it is hard to get a clear sense of exactly what the theory does and does not say. Certainly stellar and cosmic evolution, neither of which experience anything resembling natural selection, bear little resemblance to the evolution of species, and yet they share the label "evolution." But even the more narrow, purely biological evolution that occupied Darwin is complex and layered.

Biological evolution, as we understand it today, is like a digital photo composed of tiny square pixels that are normally invisible. If you zoom in too close on a digital photo, all you can see are square pixels, which look nothing like the picture. Zoom in on biological evolution and you encounter a disturbing amount of death and destruction. The majority of the offspring of many organisms simply don't make it to adulthood. Predators kill the slow and stupid; disease and bad luck wipe out many of the rest. To get one robust, fit animal across the finish line to procreation requires that nature start the race with ten animals, nine of which are doomed to a senseless death. That happy goldfinch I am watching now on my feeder has dead relatives scattered throughout my woods. This is survival of the fittest stripped of any charm or romance: the pristine wilderness where the delightful survivors cavort is littered with the bones of the less fortunate. Many of them died painfully and tragically.

Zoom out and look at evolution from farther away, and we see entire species going extinct. The fossil record—all those fascinating bones that attract kids in the science museum—is one long story of *failure*. Dinosaurs couldn't manage climate change effectively, so they went extinct, making room for mammals to rise to dominance. The dodos went extinct in the seventeenth century, unable to handle the arrival of "civilization" on their island homes. "Nature," wrote Alfred Lord Tennyson in *In Memoriam,* is "red in tooth and claw," challenging the pre-Darwinian vision of nature as a sunny meadow full of butterflies, songbirds, and lovers with picnic baskets.

Zoom all the way out, however, to the scale where ape-men are being steadily promoted and fish are scrambling onto tidal flats, and the picture gains some charm. Evolution from this vantage point looks rather glorious, working patiently over millennia to turn sponges into people and a few simple life-forms into the rich diversity that makes the world so interesting. The blood oozing in the picture up close is invisible from far away.

A deep paradox exists here between the *product* of evolution and the *process*. Most of us value life more than nonlife, complex life more than simple life, conscious life more than unconscious life, and people more than other animals. We think nothing about bulldozing ant colonies to make room for our houses; we use herbicides and pesticides with impunity, concerned only about whether they make our dogs sick. We kill mice that come indoors, and we drive owls to extinction. Most of us are fine with slaughtering cattle to make hamburgers and shooting monkeys that get too violent in the zoo. Our concerns are, first and foremost, for ourselves, our families, our communities, and our species, in that order. Those with broader concerns typically have to become a public spectacle to get heard, chaining themselves to trees or lying down in front of bulldozers.

This hierarchy of values has implications for human behavior. It also shapes the way we view evolution. If humans are more valuable than simpler life-forms, then evolution produces *value* over the course of time. This mitigates and, in the eyes of many, justifies the continual bloodletting associated with the evolutionary process. You have to break an egg to make an omelet. Or, as the more philosophically inclined might put it, the ends justify the means.

IS THERE AN END IN SIGHT?

Competition for limited resources, said Darwin, leads to improved competitors. If only the fittest survive to reproduce, then the next generation will be fitter. Darwin developed his theory to explain how species adapt in nature, but the basic idea clearly had broader applications. Any competition that consistently eliminates the weak and advances the best will produce a superior final product. Think of the Olympic athletes with their gold medals. How many lesser athletes were eliminated in their long climb to the top?

Social Darwinism is the idea that selection processes can work on different entities or "social units." People can compete; but so can teams. Towns can compete with each other to attract businesses. Ethnic groups can compete. Corporations compete for consumer dollars. Entire countries compete in everything from the Olympics to the occasional violent conflict. Each one of these social units has its own arena of competition and specific fitness criteria. If fitness is going to improve—a desirable goal—then the stronger players must be able to defeat the weaker ones. The defeat of the weak is the downside, the price paid for the generation of excellence. In the Olympics, the defeats entail personal heartbreak and even humiliation on national television. At the corporate level defeats result in companies going bankrupt and people losing their jobs. Globally they can mean war, with thousands of people losing their lives. In Darwin's theory they mean that some organisms succeed at producing offspring while others fail, often because they die before adulthood. And just as Darwinian bloodletting seemed necessary to enable a process leading to human beings—a necessary evil, so to speak—so every selection process justifies a bit of collateral carnage in the service of something larger.

In the decades after Darwin published his theory, dramatically different agendas invoked his theory as a rationale to justify various ideologies. If "survival of the fittest" was indeed a scientifically established vehicle for "progress," then why restrict it to the production of species? Why not use it, for example, to selectively weed out unfit humans in order to improve the human race? If less fit humans were sterilized by the stronger stock, then wouldn't the human race be stronger? Or perhaps unfit humans should simply be destroyed, suggested the Nazis.

In the marketplace, why not allow strong companies to run roughshod over those less able to compete? If companies compete without regulations, so the strong can drive the weak into bankruptcy, then the surviving companies will be stronger, the economy more productive, and we'll all have better iPods. And what about nations? Why not allow strong nations—with more "fit" societies—to overrun and absorb the less fit? Are not strong nations, with the various superiorities that give them their strength, to be preferred to weak ones? Who can look at the happy Canadians and not conclude that their way of life should be forcibly imposed on the poor Haitians?

Capitalists, nationalists, and racists, of course, promote agendas of self-interest and appeal to whatever rationale seems most helpful. Few of them are interested in any progress other than their own. And none of them are or were inspired by Darwin, for they have been around for ages. Two millennia before Darwin, for example, Plato championed selective breeding of humans as a way to increase the fitness of the race. His fellow Greek, Thrasymachus, preached that "might makes right," justifying the strong trampling the weak as a way to achieve more powerful political structures. The ancient Hebrews, in a campaign of reverse anti-Semitism, thought it appropriate to slaughter the men, women, children, infants, sheep, camels, donkeys, and cattle of the Amalekites, to prevent contamination of their superior religion.[4] You can't have Hebrew cows mating with pagan bulls. Such examples illustrate the countless ways that strength and fitness could be promoted through subordination of the less fit—all without any help from Darwin.

Once Darwin's theory appeared, however, Spencer and like-minded political pundits immediately adapted it to rationalize the crushing of the weak by the strong. Disturbing philosophies of self-interest thus acquired a gloss of scientific respectability, making them even more pernicious. Naïve but horrified biologists tried unsuccessfully to argue that evolution by natural selection was simply a *description* of a historical process, making no moral judgments about the ethics or integrity of this process. The historical fact that volcanoes spew lava over villagers hardly provides license for people to proactively mimic nature with moral impunity. But evolutionary theory provided an extraordinary new worldview that was especially seductive to self-congratulatory Europeans, already convinced that human history was best understood as a steady advance to the exalted plateau on which they found themselves. And much of this plateau, of course, rested on the blood, sweat, and tears of conquered peoples.

Darwinism, for better or worse, but mainly worse, has been continually attached to agendas that have nothing to do with the "origin of species." Right or wrong, but mainly wrong, Darwinism has always looked much larger than biology. And today the opposition to evolution from Christians is driven by a conviction that Darwin's theory undermines traditional values and opens doors to assorted evils. This conviction, although often poorly articulated, has ample historical precedent and should be taken seriously. The same naive and horrified biologists, of course, continue to lament this misapplication of the theory and accuse Darwin's critics of muddled thinking. But the truth is that Darwinism emerged in a socially complex milieu and has been socially embedded ever since. To understand the enduring intensity of America's reaction to Darwinism, we must acknowledge the significance of this history, not dismiss it as a trivial aberration. There is nothing new, uniquely American, or pathologically religious about seeing more in Darwinism than a simple theory to explain the origin of species by means of natural selection.

BLESSED ARE THE POOR—OR NOT

The first and most significant of the many Darwinian social agendas is one that preceded Darwin and played a role in his development of the theory. Darwin's England was caught in a struggle involving the social order. At the top, royalty claimed a divine right to their power. Privileged clergy were protected and paid by the state. There were lords with historical titles and wealthy landowners whose socioeconomic status had dubious origins in the distant past. And there were working classes and unemployed poor. This hierarchy constituted a well-defined social order that the privileged upper classes wanted to protect. But the order was everywhere under attack. Secularists blasted the entrenched power of the clergy. Reformers blasted the persistence of questionably obtained historical affluence and influence, passed down from one undeserving generation to the next. The poor rioted and assaulted the bastions of power, demanding more opportunity.

The problem of the poor was especially vexing. They tended to have larger families and were moving from rural England into the cities to work in the new factories. The industrializing cities were growing crowded and dirty and developing concentrations of these poorer classes, who were demanding attention. Their poverty, living conditions, and poor education made them susceptible to illness, criminality, drunkenness, and other vices. Those who took pity on them demanded housing, hospitals, asylums, education, and laws to protect children from abusive labor practices.

Programs to support the poor, however, inevitably lead to an increase in their numbers. Give them food and fewer will starve; give them medicine and disease will be checked; employ and educate them and they will be less likely to kill each other. Unchecked populations, unfortunately, increase exponentially: one million leads to two million then to four, then eight, sixteen, and so on. If the poor flourish, their numbers will rise faster than the resources necessary to sustain them, leading inevitably to a disastrous imbalance that will ultimately be corrected by widespread starvation. By this logic, programs to support the poor were clearly misguided. Better for half the population to starve when that number is one million, rather than when it is sixteen.

An Anglican clergyman named Thomas Malthus worked out this morbid mathematics in his widely read *An Essay on the Principle of Population,* published in 1798. "The power of population," he wrote, "is so superior to the power of the earth to produce subsistence for man, that premature death must in some shape or other visit the human race." Fortunately, death had a great many conscripts—"extermination, sickly seasons, epidemics, pestilence, and plague"—but should these front-line soldiers prove inadequate to keep overpopulation at bay, "gigantic inevitable famine stalks in the rear, and with one mighty blow levels the population with the food of the world."[5]

Darwin encountered Malthus's essay in 1838 while working on his theory. He recognized that Malthus's insight—unchecked population growth will outstrip increases in the food supply—applied to *all* species, not just humans. Therefore, since most populations are stable, there must be widespread competition for the limited resources. The fittest were winning this competition; the unfit were being weeded out, "selected" by nature for removal.

Incorporated into Darwin's theory, Malthus's principle was promoted from a depressing socioeconomic insight to full partner in the grand creative process that had sponges competing to see who could be the first to turn into a supermodel. Famine and pestilence went upscale, joining chisels and sandpaper as tools that create through destruction. Defenders of the status quo, in love with the idea that their exalted status derived from their competitive prowess, had been accused of being heartless and uncompassionate. They now leaped enthusiastically onto this shiny new Darwinian bandwagon, arguing that it was unnatural and ultimately cruel to enable any swelling of the ranks of the poor. Do nothing and let nature take its course, unless the idea of mass starvation is somehow attractive to you.

Herbert Spencer, who turned the phrase "survival of the fittest" into a household term, mocked the liberal reformers lobbying on behalf of the poor: "'They have no work,' you say. Say rather that they either refuse work or quickly turn themselves out of it. They are simply good-for-nothings, who in one way or other live on the good-for-somethings."[6] Quoting the Bible—"if any would not work neither should he eat"—Spencer argued that it was natural that "a creature not energetic enough to maintain itself must die."[7]

As for Darwin, he barely recognized his theory draped in such dark cloth. Perhaps because of his experience with his daughter Annie's death or perhaps because of the Christian charity he retained throughout his life—a charity he practiced through his family church even after he stopped attending—he was never personally able to get past the simple conviction that people should help each other, even if it meant tolerating legislation that taxed the productive members of society to provide support for the so-called "good-for-nothings." Nevertheless, he struggled with the tensions between his personal feelings and the broader implications of his theory.

EUGENICS

Leaving the poor to their own devices was a *passive* strategy to ensure that productive societies did not become diluted with useless, stupid, or otherwise defective people. Natural selection would do the dirty work as long as nothing—like misguided pie-in-the-sky liberals and their social reform agendas—interfered. But natural selection was slow and, despite the intelligence implied by the word *selection,* the process was really little more than a crapshoot with slightly loaded dice. Breeders, for example, could move tulips and dogs along the happy road of progress much faster using *artificial* rather than *natural* selection. Why not assist Mother Nature by inserting a bit of intelligence?

It was Darwin's cousin, Francis Galton, who suggested that the pestilential growth of the lower classes required something more aggressive than unaided natural selection. Even without social assistance, the downtown slums were filling up with lunatics and criminals, the result of unchecked procreation. In contrast, the superior residents of the uptown penthouses were having fewer children, sensibly moderating their procreation. For anyone who could do the math, the social trajectory looked grim. The human race, at least in England, was deteriorating.

Galton's solution was simple: encourage the more fit members of society to have more children, just as better cattle are bred by mating the stronger members of that species. In an 1865 article in *Macmillan's Magazine* titled "Hereditary Talent and Character," he outlined his vision for the production of a superrace of humans:

> If a twentieth part of the cost and pains were spent in measures for the improvement of the human race that is spent on the improvement of the breed of horses and cattle, what a galaxy of genius might we not create! We might introduce prophets and high priests of civilization into the world, as surely as we can propagate idiots by mating *cretins.*[8]

Galton, like most everyone at the time, including cousin Charles, was deeply racist. His classic treatise, *Hereditary Genius,* is filled with the most natural and straightforward analysis of the quality of England's public figures as well as the races of the world. The chapter titled "The Comparative Worth of Different Races" offers a sobering portrait of nineteenth-century Victorian elitism.[9]

The self-congratulatory Victorian obsession with progress, coupled to the belief that evolution produced "higher" creatures from "lower" ones, had nearly everyone convinced that the various human "races" could be ranked. Not surprisingly, the scale had white Europeans at the top, reciting poetry while eating cooked foods off china with knives and forks rather than plucking bananas from trees like lower primates. If the lower races became extinct, that would represent progress. Galton and his followers were quite animated about the prospects of breeding the best representatives of the most advanced culture to create a superrace. Galton coined the term *eugenics,* meaning "best born," to describe his program. It was a program that soon found itself shrouded in a dark, sinister fog.

THREE GENERATIONS OF IMBECILES

The eugenics movement became popular in most of Europe, Canada, and the United States. In the United States an influential study by a New York social reformer, Richard Dugdale, traced the "Jukes" family through five generations, establishing that most of the 709 relatives examined were "criminals, prostitutes, or destitute."[10] Convinced that such defects were hereditary, reformers enthusiastically promoted legislation to forcibly sterilize such defectives or otherwise prevent them from breeding.

Government offices sprang up to create eugenics policies and track progress. In the United States this was done through the Eugenics Record Office, created in 1910 with donations from wealthy American industrialists. Between 1900 and 1935, thirty-two states enacted laws permitting forced sterilization of defective humans. More than sixty thousand people were sterilized for defects ranging from "feeblemindedness" to epilepsy. Virtually every state in the United States and every country in Europe had some kind of a program to prevent defective humans from passing on their defects.

In 1914 the Eugenics Record Office developed a proposal to sterilize one-tenth of the population of every generation,[11] until fifteen million people had been sterilized. The sarcastic journalist who became famous covering the Scopes trial, H. L. Mencken, suggested that all the sharecroppers in the South should be sterilized.

The public schools taught children to think hard about choosing marriage partners and warned about the drain on society caused by defective humans. In the textbook from which John Scopes was accused of teaching evolution, author George W. Hunter outlines the social disaster of the infamous Jukes family, which produced "24 confirmed drunkards, 3 epileptics, and 143 feebleminded" as well as 33 who were "sexually immoral." Such families, the children read, were "parasites," spreading "disease, immorality, and crime to all parts of the country."[12] In a chapter titled "Heredity and Variation," Hunter continues:

> The cost to society of such families is very severe. Just as certain animals or plants become parasitic on other plants or animals, these families have become parasitic on society. They not only do harm to others by corrupting, stealing, or spreading disease, but they are actually protected and cared for by the state out of public money. Largely for them the poorhouse and the asylum exist. They take from society, but they give nothing in return. They are true parasites.[13]

In both the United States and Britain Protestant clergy floated proposals that would have required certificates of "eugenic fitness" before getting approval for a church wedding.[14] There was no room in their inns for feebleminded children. Only the Catholic Church seemed consistently concerned about these proposals for governmental meddling in human reproduction.

The U.S. Supreme Court, with a lone dissenting voice, ruled in 1927 that mandatory sterilization was constitutional for patients in mental institutions. In a landmark case, plaintiff Carrie Buck was forcibly sterilized for being "feebleminded." At the time she was a patient at the Virginia State Colony for Epileptics and Feebleminded. Her mother, Emma, had also been accused of being "feebleminded," as was her daughter, Vivian, who was sterilized as a child.

Invoking the "public welfare," in 1927 Justice Oliver Wendell Holmes wrote for the court that such "manifestly unfit" people should be prevented from breeding, rather than "waiting to execute degenerate offspring." Claiming that Carrie, her mother, and her daughter were all "feebleminded," the court ruled that the public good was served and the Constitution upheld by forcibly sterilizing Carrie Buck. "Three generations of imbeciles," wrote Holmes in a chilling conclusion, "is enough."

Eventually paroled, Carrie Buck was an avid reader until she died in 1983; the case against her "feeblemindedness" was undermined when it was discovered that a relative of her adopted family had raped her. Carrie had been committed to hide the rape and protect the family's "good name." Her "feebleminded" daughter, Vivian, died at age eight, leaving behind an academic record of modest success, including being on the honor roll.

THE FINAL SOLUTION

Eugenics took a sinister turn in Europe, especially Germany, and subsequently fell so far from grace that it became a concept from which politically savvy people would flee. Galton, we will recall, originally made the benign suggestion that the quality of the human race would be improved if the "fittest" members of society had more children. Even though Galton held a low opinion of the poor, he understood them to be a part of the species that included him and his "fit" colleagues. This diverse group, with its great variation in fitness, would be improved if the less fit had fewer children and the fitter folks had more children. This form of "positive" eugenics is still practiced through the marketing of "superior" eggs and sperm. (I have in my files, for example, a request from a sperm bank for a sample. Apparently I have passed some criteria for "fitness.")

Galton's eugenics strategies would work, at least in principle, on any group of organisms. Applied to a human group—Caucasians, Asians, baseball fans, Trekkies, or even the readers of this book—selective breeding of the "fittest" members would improve group averages. But this was not the only way to understand eugenics.

Nineteenth-century Europeans, after three centuries of global exploration and empire building, were only too aware of the different human "races." And there was general agreement that racial groups were fundamentally unequal, as both the European and American legacy regarding slavery makes painfully clear. It was only natural, argued some, to view biological competition as between entire *races,* rather than between the individual *members* of a specific race. Extending eugenics into the realm of race relations seemed entirely reasonable and logical, based on what they knew at the time. If Malthus was correct that Caucasian England was better off with fewer poor people and that reducing their number should be the goal of social policy, then it followed that the human race as a whole, or any subgroup in it, was better off if the populations of the weaker segments were reduced.

The cold trajectory of this logic is all too easy to see. In its more benign incarnations it resulted in restrictions on immigration. In 1924, for example, the U.S. Congress passed laws restricting immigration from countries and ethnic groups perceived as inferior. Such laws had a glossy scientific veneer, and racist politicians took comfort in the sophistication and wisdom of policies informed by the best science of the day.

More sobering developments in Germany led to a national program of extermination of groups perceived to be inferior. Hitler and his Third Reich viewed Jews, gypsies, Poles, and homosexuals as inferior. Ernst Haeckel nudged the racism of the Third Reich along its malignant road by suggesting that the various human races were like stages in the embryonic development of a fetus. He arrayed the various human races along a ladder with subhuman primates at the bottom and Aryan supermen like him at the top. Black Africans and Tasmanians, in his scheme, were closer to animals than to the advanced European races. It would be hard to imagine a more dangerous articulation of racism than Haeckel's. Not surprisingly, the Nazis eagerly embraced his ideas. Eager to rationalize their calculated genocide to a well-educated and culturally sophisticated populace, the Nazis invoked science whenever it served their interests and ignored it when it did not.

"If you draw a sharp boundary," Haeckel wrote in a popular book published in 1868, "you must draw it between the most highly developed civilized people on the one hand and the crudest primitive people on the other, and unite the latter with the animals."[15] The book contained sketches illustrating Haeckel's imaginative "ranking" based on the shapes of the heads of the various races.

How shocking it is today to acknowledge that virtually every educated person in Western culture at the time, on both sides of the Atlantic, shared Haeckel's ideas. Countless atrocities around the globe were rationalized by the belief that superior races were improving the planet by exterminating defective elements. This expressed itself in a variety of imperial attitudes toward non-Western peoples, lifestyles, religious practices, and ethics. The particular atrocities, of course, were not inspired by this version of social Darwinism, but there can be little doubt that such viewpoints muted voices that would otherwise have been raised in protest.

Empire-building imperialists invoked social Darwinism to rationalize colonial subordination and even organized slaughter of conquered peoples. The enslavement of blacks, the destruction of Native Americans, and the genocidal treatment of aboriginal tribes in Australia were defended as part of a grand Darwinian project to advance humanity. Joseph Le Conte, a respected geologist and president of the American Association for the Advancement of Science, addressed this issue in *The Race Problem in the South,* published in 1892. Le Conte argued that the docile character of the Negroes made them appropriate for enslavement; for races like the "redskin," however, who were more specialized and thus less flexible, "extermination is unavoidable."[16]

Well-meaning Christians, alas, believed they had license to abuse both their horses and their black servants (formerly slaves), since the best science of the day taught that neither was fully human. Racist theories like those of Haeckel produced a moral fog that made it hard for even Christians to show compassion and charity to those in need, if they fell outside certain boundaries. The most frightening incarnation of social Darwinism, of course, was Hitler's eugenics program, which eventually sent twelve million "defective" humans, half of them Jews, to various execution chambers. Nazi anti-Semitism, of course, did not originate with Darwin. In fact, there is more blame to be laid at the feet of Martin Luther than Charles Darwin. Luther had described Jews as "poisonous envenomed worms" and encouraged Christians to destroy them, inaugurating hostilities that continued unabated into the twentieth century.[17]

But German racial politics needed scientific, not religious, rationale and looked eagerly to Darwinism. Many Nazis were, to be sure, dullards and thugs easily manipulated by Hitler, with his peculiar malignant genius. Hitler certainly didn't need Darwin to help him abuse Jews. But there were many sophisticated Nazis, teaching at universities and holding high posts in the government and state churches. They needed something more than the anti-Semitic rants of their deranged führer to get behind the "final solution." Stung by their humiliation after losing World War I, Germans wanted nothing so much as to regain the glory of their past. If eliminating defective elements within their borders could accomplish this, then they were on board. And so much the better that there was a scientific rationale for this project.

I hasten to point out that the connection between Darwinism and movements like Nazism is not causal, as some shrill anti-evolutionary pundits like Ann Coulter claim.[18] Aryan Germans were not happily playing soccer and eating bagels with Jewish Germans until Darwin convinced them this was a bad idea. The connection is, rather, one of *rhetoric* and *rationalization*. It is rhetorical in the sense that dumb ideas play better when dressed in fine clothes. It sounds better to promote "cleansing the human race" than "killing people you don't like," a distinction of no value to Jews en route to Auschwitz. It is rationalization in the sense that conclusions already embraced rest easier on one's conscience if supported by some thread of rational argument, no matter how thin. The relevance of these considerations, however, is not that Darwinism leads somehow to dreadful social policies. The point is, rather, that Darwinism has been, for all of its short life, hanging out in some rather terrible company and has now got a reputation.

There is no shortage of creative rationalizations of Nazi anti-Semitism; for our purposes here one example will suffice. And although this example highlights the victimization of Jews, nearly identical arguments were applied to Negroes, Native Americans, and just about any group outside of Caucasian Europeans.

The following argument comes from Alfred Kirchhoff, a geographer at the University of Leipzig, who posthumously published *Darwinism Applied to Peoples and States* in 1910. Kirchhoff, like many evolutionists, believed that morality had evolved along with the physical and mental structures of organisms. Obviously, primitive life-forms had no morality. What exactly, could a sponge do that was *wrong*? More complex animals, like primates, had a simple morality. "Lower" human races, such as blacks and Native Americans, had a more developed morality. The higher races, which for Kirchhoff meant Europeans, had the most advanced morality. The "average" morality of the entire human race was thus "lowered" by the presence of morally inferior subgroups, just as the performance of an orchestra is compromised by the presence of a few bad musicians. So in a breathtaking application of this logic, an argument was developed that morality would actually *increase* if the morally advanced European races eliminated the morally underdeveloped races. Invoking Darwin, Kirchhoff defended this genocidal agenda, calling it the "righteousness of the struggle for existence." This struggle would lead to "the extermination of the crude, immoral hordes." The diversity of races and the resulting struggle were necessary for the "progress of humanity."[19]

As we've seen repeatedly in our discussion of Darwin and the nineteenth century, educated Europeans were marching in lockstep behind the pied piper of progress. Progress was now a moral crusade, and policies perceived as progressive needed little additional justification. *Might,* as an enabler of progress, slowly, imperceptibly, turned into *right,* in the eyes of far too many of Europe's leading lights. Subtle statements and innuendos, in textbooks like that used by John Scopes, acclimated schoolchildren to this mind-set. Less subtle, deeply political messages appeared in places like *Mein Kampf*, where Hitler waxed eloquent about the triumph of the strong, calling it an "iron law of necessity," justified as the "right of victory of the best." Note the value judgment implied by the word "best." "Whoever will not fight in this world of eternal struggle," Hitler wrote in language eerily reminiscent of Darwin's explanation of natural selection, "does not deserve to live."[20]

A small library of books could be assembled rationalizing the sinister ideas that became incarnate at Auschwitz and Dachau. Educated Germans designed efficient killing machines, over which trained medical personnel presided, for the purpose of advancing the human race through the destruction of the weak. This is a tragic chapter in German history that scholars are still trying to understand. But one thing is crystal clear: the Holocaust would have happened with or without Charles Darwin. There can be no doubt, however, that the Nazi campaign against the Jews was assisted via *rhetoric* and *rationalization* with arguments from social Darwinism.[21]

UNHAPPY BEDFELLOWS

The connection between biological and social Darwinism is complex and troubling, and perhaps even suspicious, but there is no denying that it has always been there, even before evolutionary theory became known as "Darwinism." The arguments and even some of the practices are still in play. Most sperm banks take in account eugenic considerations. Parents routinely test for birth defects, and "defective" embryos are often aborted. William Shockley, who won a Nobel Prize for physics in 1956, used his fame as a platform to warn humanity about "the genetic deterioration of the human race through lack of elimination of the least fit as the basis of continuing evolution."[22] In 1994 two prominent social scientists, Richard Herrnstein and Charles Murray, reignited the controversy with their book *The Bell Curve: Intelligence and Class Structure in American Life*. The authors updated eugenic concerns identical to those that worried Galton: "higher fertility and faster generational cycle among the less intelligent." This has dire social consequences. "Something worth worrying about," they warn, "is happening to the cognitive capacities of the country."[23]

Storms of occasionally violent protest greeted the eugenic agendas of Shockley and *The Bell Curve*. Few ideas upset contemporary sensibilities more than the suggestion that intelligence varies by race. Its association with that idea gives Darwin's theory a stench that many find unbearable. And there are other, equally troubling connections drawn between evolution and unpopular ideas, including philandering, infanticide, violence, and rape. The Harvard linguist Steven Pinker invokes Darwinian principles to explain infanticide, suggesting that killing one's newborn should not be viewed with the same seriousness as killing one's child later in life.[24] The authors of *A Natural History of Rape* invoke Darwin to explain that rape is a consequence of "men's evolved machinery for obtaining a high number of mates."[25] Connecting evolution to racism, rape, infanticide, philandering, and so on makes many people very nervous.

Thoughtful evolutionists hasten to point out that no necessary connection exists between biological evolution, which provides *descriptive* explanations of how nature works, and social Darwinism, which suggests *prescriptive* guidelines for how society *should* behave. It is far from obvious that eugenics, unbridled capitalism, relaxed attitudes about infanticide, or rampant militarism is implied by the theory that species originate through natural selection.

Let us suppose, for the sake of argument, that such extensions are warranted, perhaps in the service of some "greater good." We immediately face a host of ambiguities. How do we actually apply Darwinian principles to social behaviors? Consider the relatively benign world of capitalism, with iPods, Toyotas, and dishwashers. Applying Darwin's "survival of the fittest" to resolve the dispute over Apple's aggressive business practices, for example, is far from straightforward.

For starters, since we are making a *moral* judgment about actions in the free market, we must decide what it is that *should* have its "fitness" protected. Is it the *products* competing in the portable music space? Are they the units of Darwinian selection? Do we want to enable the development of the best and most fit music players and best online music stores? If so, then we should not allow Apple to artificially enhance the market share of either the iPod or the iTunes store by linking them in a way that makes it harder for other products to get into this market space. Such a practice would be anticompetitive and non-Darwinian. But what if, instead of using competition to enhance the fitness of *products,* we look instead at ways to use competition to enhance the fitness of the *companies* that make the products? Maybe the "unit of selection" is Apple Corporation, not the iPod. Certainly when Andrew Carnegie appealed to Darwinism to justify aggressive business practices, he was more interested in the fitness of the corporations than the products they produced. Under this interpretation we should want the *company* to be stronger, and we should allow Apple Computer do whatever it wishes to grow its market share and profits. What should disappear in this competition are not competitors' products but the competitors themselves.

Applying social Darwinism to society creates the same problems. Is the unit of selection the *individual,* for example, or the *society*? Is the competition between *people* or between *countries*? Is the competition *military* or *economic*? And how is success defined? In the United States we hear a lot about the gross national product and how important it is for that to become larger. In dramatic contrast to this, tiny Bhutan calculates a "gross national happiness" and works to increase that index. Which is the better measure of Darwinian fitness?[26]

Social Darwinism turns out to be almost useless when you actually try to do something with it. Many ideologically driven decisions have to be made before you can even apply it. As a result, it ends up being little more than a bogus appeal to science to rationalize an agenda already embraced for other reasons.

CONCLUSIONS

Despite nonstop critique by philosophically sophisticated evolutionists, pundits continue to find, within the science of biological evolution, justification for controversial moral stances on an array of social problems. Right or wrong, but mainly wrong, evolution continues to be connected to far more than the historical origin of species. And these connections exacerbate whatever concerns people might have about whether evolution is actually "true." Believing something is false is much easier when you desperately want it to be false.

These associations are problematic in the context of America's current controversy over evolution. Half the population of America thinks evolution is simply not true. For this vast constituency, God created the species individually; they did not evolve by natural selection or any other method. The evidence is unconvincing, the religious problems overwhelming, and the idea that a sponge could turn into a person is ridiculous. Furthermore, there are noisy "creation scientists" and "intelligent design theorists" highlighting the problems with evolution and offering simplistic alternatives that satisfy the limited curiosity of most Americans about origins.

The morally complex baggage carried by evolution hampers its acceptance. Even if evolutionary theory were true, why would anyone *want* to believe a theory that rationalizes Nazism, infanticide, and rape? The theory's supposed "explanation" of these horrors represents for its detractors further evidence that the theory is really just a secular myth, undermining morality, condoning evil, and destroying religion.

Curiously, surveys of evolution by its many eloquent advocates gloss over social Darwinism as little more than an historical aberration. In the companion volume to the seven-part PBS series on evolution, science writer Carl Zimmer draws no connection whatsoever between evolution and social Darwinism. It appears only in a discussion of religious objections, where it is dismissed as "scientifically baseless."[27] According to Zimmer, people like William Jennings Bryan, the lawyer who prosecuted John Scopes and who despised Darwinism for its apparent evil implications, were simply confused.[28] Presumably the biologists who wrote the textbooks used in the high schools at the time were similarly confused, although Zimmer makes no mention of them. Niles Eldredge, curator at the American Museum of Natural History, comments in *Darwin: Discovering the Tree of Life* only that there is no "neat one-on-one correspondence between evolution and any single system of ethics."[29] Eugenie C. Scott, who heads up the National Center for Science Education, barely mentions it in her encyclopedic *Evolution vs. Creationism*.[30] Ernst Mayr, who until his death in 2005 was the dean of American evolutionists, relegates social Darwinism to a two-line entry in the glossary of his authoritative *What Evolution Is*. The text contains no discussion of it at all.[31]

These recent treatments contrast sharply with the history of evolution. Are social Darwinism and evolutionary theory really as unrelated as today's champions of evolution claim? How, then, did their predecessors get it so thoroughly wrong? It took Nazism, apparently, to deflate the eugenics balloon and two world wars to silence the loudest of the "might makes right" enthusiasts. Yet now we are told that these connections should never have been made and that they derived from "confusion."

Is it not disingenuous for evolutionists to pretend that these historical connections are aberrations? How many times do we see John Scopes held up as a martyr for the noble cause of teaching schoolchildren the truth, and yet we never read a word of criticism about the racism in the text he used? William Jennings Bryan continues to be ridiculed for thinking that social Darwinism contributed to World War I, but American eugenicists who, in the name of Darwin, sterilized thousands of people against their will have strangely disappeared from history. Darwin's dark companions are being written out of history, like characters in George Orwell's novel *1984*.

Popular books disputing evolution, not surprisingly, give plenty of space to Darwin's dark companions. Of course, the goal of this propaganda is to nurture revulsion in their readers against evolution and convince them that it is truly a Satanic theory, as Ken Ham and Henry Morris claim. Ham's book *The Lie: Evolution* contains a chapter titled "The Evils of Evolution," which opens with a drawing showing evolution as the literal foundation of lawlessness, homosexuality, pornography, and abortion. Parallel section headings in the chapter link the following to evolution: Nazism, racism, drugs, abortion, business methods, and male chauvinism.[32]

Conservatives, by tradition and perhaps by definition, have always lamented the direction and pace of social change. From Plato and Socrates decrying the ruffians of ancient Greece to anti-evolutionary crusaders warning about the misguided youth of today, there has always been hand-wringing about change. And every generation of conservatives needs a scapegoat against which to rally the faithful. Tragically, Darwin plays that role today, as the most preposterous charges are leveled against him. In a slickly produced DVD from Coral Ridge Ministries, the late D. James Kennedy blames Darwin for everything from the Holocaust to the shootings at the Columbine high school.[33] "If evolution is true," Kennedy writes in the foreword to a companion book, "then we are simply the product of time and chance, and there is no morality and no intrinsic worth to human life."[34]

Ann Coulter, Lee Strobel, and other anti-evolutionary culture warriors join Kennedy on the DVD in a disturbing and appalling piece of propaganda dramatically at odds with contemporary scholarship. They can be forgiven, perhaps, since they are neither scientists nor historians of science. But they are so thoroughly and completely wrong that it is hard to imagine that they believe their own rhetoric. Perhaps what they believe themselves is not that important. But, unfortunately, millions of Americans are listening.

These sinister portrayals of Darwin and his dangerous theory frighten millions of ordinary people. True or false, who wants a theory that destroys all that is noble and good about being human? And who would want their taxes supporting the teaching of such dreadful falsehoods to their children? So, when Darwin's dangerous idea began to show up in the public schools, there was an immediate reaction. From Dayton, Tennessee, where John Scopes stood trial for teaching evolution, to Dover, Pennsylvania, where a local school board tried to wriggle intelligent design into the curriculum, evolution has had a nearly permanent home in America's courtrooms.

CHAPTER 4

**THE NEVER ENDING
CLOSING ARGUMENT**

John Scopes, by all accounts, was a nice guy. He taught a variety of subjects, including biology, in the local high school in Dayton, Tennessee. One fateful day, in the middle of a lesson explaining Darwin's theory of evolution to his students, the local sheriff dropped by. The sheriff, with a couple of other prominent local citizens including a clergyman, stood ominously at the rear of the classroom, listening to Scopes's explanation of evolution. Put off by the strange and uninvited visitors, Scopes did the best he could to maintain normalcy in the class and continue the lesson:

> Darwin's theory tells us that man evolved from a lower order of animals: from the first wiggly protozoa here in the sea, to the ape, and finally to man. And some of you fellas out there are probably gonna say that's why some of us act like monkeys ...

The sheriff interrupted Scopes and made a great show of verifying his identity, although they had known each other for years. Reading from a paper, he informed the young biology teacher that he had broken the law that forbade "any teacher of the public school to teach any theory that denies the creation of man as taught in the Bible and to teach instead that man has descended from a lower order of animals." The sheriff arrested Scopes and took him to the local jail. His crime was the teaching of evolution. While he was in jail, the local citizens, seemingly all mean-spirited fundamentalist Christians, burned Scopes in effigy and sang about hanging him "from the sour apple tree."

The subsequent trial was a great media event. The famous politician William Jennings Bryan represented the state of Tennessee, and the equally famous agnostic lawyer Clarence Darrow represented John Scopes. Darrow, representing both Scopes and evolution, humiliated both Bryan and the anti-evolutionary forces. Bryan dropped dead immediately after the verdict. And the anti-evolutionary forces went into hiding.[1]

This is the famous "Monkey Trial," which, apart from Jesus's trial before Pilate, is probably the best-known legal confrontation in history. There is one problem, however. The events outlined above, although at least vaguely familiar to most educated Americans, did not occur.

WILL THE REAL JOHN SCOPES PLEASE STAND UP?

The Scopes trial was a definitive moment in American history. It captured the nation's attention like no trial before or since and now sits in the background of all confrontations between creation and evolution.

The trial contains both more and less than meets the eye. It is less in the sense that the trial was really just a show, and none of the players were interested in the trial itself. The version above, which most people recall as the history of the trial, comes from *Inherit the Wind,* a movie inspired by and *loosely* based on the trial. The play of the same name had also played widely across America before being brought to the big screen. But the Scopes trial is also more than meets the eye, incarnating the inevitable and cataclysmic confrontation of two cultural groups. A new secular America was emerging, with a theory assaulting the traditional story of creation. Traditionally religious Americans opposed secularization, of course. And they were uneasy about a theory that rationalized all manner of social evils from eugenic sterilization of the feebleminded to the invasion of weak countries by the strong.

The legend of the Scopes trial, in popular mythology, makes many conservatives see red. The national reporting of the trial was dominated by H. L. Mencken of the *Baltimore Sun,* probably the most sarcastic journalist who ever worked in the English language. Mencken hated Bryan, the South, small-town America, and fundamentalists, which he saw as something of a stone-age package. He loved Darrow, hyperbole, Northern liberalism, and the sound of his own typewriter. "Fundamentalists," wrote Mencken shortly after the trial, are thick in the mean streets and gas-works. They are everywhere where learning is too heavy a burden for mortal minds to carry, even the vague, pathetic learning on tap in the little red schoolhouses. They march with the Klan, with the Christian Endeavor Society, with the Junior Order of United American Mechanics, with the Epworth League, with all the Rococo bands that poor and unhappy folk organize to bring some new light of purpose into their lives.[2]

The unofficial "history" of the Scopes trial appeared in 1931, six years after the verdict. A leading journalist wrote a lively, sensationalized, simplistic, and best-selling history of the Roaring Twenties. In true journalistic fashion the story became a confrontation between well-defined and polarized opponents, without complexities or middle ground. The loser, conservative Christianity, was led by a political dinosaur named Bryan who went extinct one week after the trial ended. The winner was twentieth-century skepticism, led by a liberal crusader on a white horse named Clarence Darrow.[3]

In giving cinematic life to this drama, *Inherit the Wind* further maligned Bryan, transforming him into a pompous buffoon named Matthew Harrison Brady, played with enthusiasm by Frederic March. The town of Dayton, which Mencken conceded was "full of charm and even beauty,"[4] became a creepy cultural backwater. The townsfolk, mainly just old-fashioned Christians, became narrow-minded bigots, hostile to progress and science. The sedate faith of the real-life Daytonians became dark and sinister and was personified in the Reverend Brown, a vindictive and totally fictional character. The unsympathetic Brown, representing the worst in anti-evolutionary bigotry, calls down God's wrath on his daughter for supporting Bertram Cates, the Scopes character. The controversial American Civil Liberties Union (ACLU) attorney, Darrow, was named William Henry Drummond and played by the beloved Spencer Tracy. *Inherit the Wind* portrays this character as lovable and heroic, a warm grandfatherly figure. This portrayal stands in stark contrast to the real-life Darrow, a notorious lawyer who had just defended two rich Chicago teenagers who had murdered fourteen-year-old Bobby Franks for kicks.

Many view America's creation–evolution controversy as a part of the war between science and religion. Through this lens, science appears to triumph over religion at Dayton. This was not, technically, the verdict, but who cares about the truth when the myth is so interesting? Strangely, however, the true story is actually less plausible than the legend.

JUST THE FACTS, PLEASE

I grew up in a small town like Dayton, Tennessee, where everybody knows everybody else, concerns are mostly local, and the rest of the world seems far away. Such small towns, rendered so faithfully in Sinclair Lewis's novels, have informal meeting places where gossip and town business are always on the agenda. In my boyhood town of Bath, New Brunswick—population one thousand—the location was Abe's barbershop, which always had way more men on its hard wooden benches than needed haircuts. In Dayton, Tennessee—population eighteen hundred—in the 1920s, the place to meet was the soda fountain in Fred Robinson's drugstore. The topic on May 4, 1925, was evolution in the public schools, an unusually weighty subject.

The conversation that led to the Scopes trial started innocently, if ominously, on Friday, March 13, 1923, when the Tennessee Senate made it illegal "to teach any theory that denies the Story of Divine Creation of man as taught in the Bible, and to teach instead that man has descended from a lower order of animal." The Tennessee law was drafted by a farmer and part-time schoolteacher named John Washington Butler, who was also the clerk of the Round Lick Association of Primitive Baptists. It was the second of many such laws that states were passing to keep Darwin's dangerous idea away from their children, who were at last staying in school past the eighth grade.

The ACLU believed the law violated the Fourteenth Amendment, forbidding states from depriving anyone, including public-school teachers like John Scopes, of "life, liberty, or property without due process of law." The Tennessee law clearly restricted what a public-school teacher could teach, and the ACLU had been looking for a case like this. About a month after the law went into effect, the ACLU started running an advertisement in a newspaper in Chattanooga, forty miles from Dayton, looking for a guinea pig who would admit to the crime of teaching evolution to the children of Tennessee. It offered to pay the legal expenses of this criminal.

Everyone knew that the trial the ACLU was orchestrating would be a big show. Probably William Jennings Bryan, whose personal anti-evolutionary crusade had inspired much of the associated legislation, would ride in on his populist bandwagon to defend the law. And certainly the ACLU would import some colorful, arrogant big shot from the North to defend evolution and ridicule those who opposed it. Interested onlookers, probably in the thousands, would pour into whatever town hosted the event.

The locals at Robinson's drugstore thought it would be good for business if all these legal enthusiasts came to Dayton. Hotels would fill up, restaurants would boom, and Robinson would sell lots of sodas. Of course, if Dayton were to host the trial, they would need to find a local criminal who had broken the law against teaching evolution. And tiny Dayton with its handful of conservative Christian high-school teachers had few candidates. But one came to mind.

John Scopes, a general science teacher who taught physics and math and coached football, was one of the few "liberals" in Dayton. He willingly embraced the role proposed by the drugstore conspirators. He had no recollection of having taught evolution, but he had filled in once for a biology teacher in a class that used a textbook, Hunter's *Civic Biology,* which did have a few pages on evolution. As it turned out, that was close enough, and Scopes was "arrested" for breaking the law against teaching theories of origins in conflict with those contained in Genesis. The crime and the arrest were total shams of course, and Scopes worried about being exposed:

> I didn't violate the law.... I never taught that evolution lesson. I skipped it. I was doing something else the day I should have taught it, and I missed the whole lesson about Darwin and never did teach it. Those kids they put on the stand couldn't remember what I taught them three months ago. They were coached by the lawyers.[5]

As expected, William Jennings Bryan, the "Great Commoner," immediately presented himself as attorney for the prosecution. Although courtroom sparring was an unfamiliar game to this lifelong politician, he was arguably the country's greatest orator, and he welcomed the opportunity to bring his anti-evolutionary cause to Dayton. Clarence Darrow, the agnostic lawyer, elbowed his way onto the defense team, looking eagerly past Scopes to Bryan, with whom he had been sparring in print for years.

Also as expected, tiny Dayton was overrun with reporters, spectators, and the occasional expert witness. The hotels filled, and Robinson sold many sodas. The *Baltimore Sun* sent H. L. Mencken, who regaled eager Northern readers with witty, sarcastic, and partially true stories about the great confrontation in Dayton, where the nineteenth century was going head-to-head with the twentieth. The *Courier Journal* of Louisville, Kentucky, ran a headline on July 21, 1925: "3,000 AT TRIAL, GET THRILL."

The "Monkey Trial," as it was known, was peculiar in many ways. The defense brought expert witnesses to testify that evolution was a mainstream biological idea and should be taught in the public schools. The court ruled their testimony inadmissible and irrelevant, since the truth of evolution was not the issue, so these experts sat on the sidelines. The law said simply that human evolution could not be taught. And there was little doubt that Scopes had indeed confessed to teaching that humans had evolved.

The trial roared to life when Darrow called Bryan as an expert witness on the Bible. The judge had frustrated the defense by ruling that its expert scientific witnesses could not testify to the truth of evolution. And its expert theological witnesses did not get to testify to the compatibility of the Bible and evolution. So Darrow resorted to a quixotic, self-aggrandizing, and ultimately brilliant legal maneuver in calling the opposing counsel as an expert witness. Bryan could have refused, of course. What advantage did he perceive in serving as an expert witness for his adversary? But Bryan was a powerful orator who could move crowds with his eloquence. He had a solid layman's familiarity with the Bible, which he could quote to great effect. And there were settings in which audiences would marvel at his command of the Bible. Unfortunately, the witness chair on the lawn outside the Dayton courtroom, where the judge moved the trial, would not prove to be such a setting. Bryan's knowledge of Scripture was almost purely devotional, and he was unfamiliar with the problems that even elementary biblical scholarship was raising.

As a flame draws a moth, the witness stand beckoned to an unprepared Bryan. Years past his prime, he found himself outmatched by the wily Darrow. Sensing Bryan's vulnerability, Darrow circled logically about him, nipping at the Great Commoner's heels with standard village atheist fare: Where did Cain get his wife? Was Jonah swallowed by a whale? Did God really make the sun stand still for Joshua? These were questions Darrow had been lobbing at Bryan for years in print, and now Darrow had him, tethered to a witness stand with the entire world watching. The audience on the lawn grew from five hundred to three thousand, energizing the actors. This was classic theater, and Darrow knew it; even the playwrights who infused so much dramatic fiction into *Inherit the Wind* could scarcely improve on the drama of the actual interrogation. Of course, all of this had nothing to do with whether John Scopes should be convicted for teaching evolution, but it was the epic struggle that put Dayton and the Scopes trial on the national map.

How much Bryan embarrassed himself on the witness stand is hard to say. He died one week after the trial ended, silencing critics uneasy about flogging a national corpse. Transcripts and eyewitness accounts, though, certainly indicate moments of muddle. When Darrow pressed him on the date of Noah's flood, Bryan hedged, saying he had not thought about it much, deferring to unnamed scholars who had written on the topic. Darrow wanted some kind of answer and pressed Bryan further: "What do you think?" he asked, to which Bryan responded, "I do not think about things I don't think about." Darrow came back with, "Do you think about things you do think about?" and Bryan responded "Well, sometimes."[6] The assembled crowd laughed, clearly *at* and not *with* the Great Commoner.

Certainly the media, dominated by the North's low opinion of all things Southern, concluded that Bryan, fundamentalism, and the creationist cause had been thoroughly humiliated. The *New York Times* called the Darrow-Bryan duel "an absurdly pathetic performance."[7] Even Tennessee papers were critical: "Darrow succeeded," wrote a Memphis paper, "in showing that Bryan knows little about the science of the world."[8]

But great men and great causes cannot be reduced to their worst moments. The fires of anti-evolutionary fundamentalism barely flickered at Dayton and soon came roaring back. As for Dayton, it built a college to honor the great hero who fell on its battlefield. William Jennings Bryan College opened for classes on September 18, 1930, in the old school where John Scopes did not break the law against teaching evolution. Bryan College is now a healthy Christian liberal arts college with thousands of graduates, one of whom is my sister, who received her mathematics degree in 1981.

POST-SCOPES

The aftermath of the Scopes trial clarifies an important theme in America's creation–evolution controversy, namely, the great divide that began to separate ordinary religious people from the educated leadership of the country. Bryan saw this only too clearly at Dayton, warning the common people who idolized him not to turn over the education of their children to an elite establishment that did not share their values. Whether evolution was right or wrong, and Bryan clearly believed it was wrong, it conflicted with the religion of most Americans. Ordinary taxpayers, argued Bryan, should be empowered to prohibit public schools from teaching their children things in conflict with what they were learning at home and in their churches. The public schools should serve common, ordinary people, not ivory-tower elites with no appreciation for traditional values. Arguments like these earned Bryan his nickname, the Great Commoner.

America's great divide over creation and evolution is a complex cultural phenomenon, largely because of the way power is distributed in America. The power divide establishes a significant distance between ordinary people, of which there are many, and elite leaders, of which there are few. If I may be forgiven some oversimplification, this can be pictured as a pyramid, with multitudes of ordinary people at the bottom supporting an increasingly smaller number of more educated leaders at the top. Anti-evolutionary sentiments are strongest at the bottom, weakest at the top.

In America's culture war over evolution the base of this pyramid wields its power to fight evolution as consumers, taxpayers, and voters. The top of the pyramid wields its power through control of the government, the courts, the universities, and the media. In that stifling hot courtroom in Dayton, the base of this pyramid was represented by the twelve jurors, all farmers, and the locals who came to watch. The tip of the pyramid was represented by Mencken, Darrow, and the expert witnesses from the universities. Bryan's greatness lay in his ability to support people like the farmers in their struggle against powerful leaders who disrespected their values.

Every time creationism has clashed with evolution, this same tension has been present. Ordinary people concerned about evolution organize on state and local levels and take their concerns to the next level, which is inevitably "higher up" on the pyramid. As concern rises upward from the base, it encounters increased opposition as it ascends farther from its base of support. In the court cases that followed Scopes, we repeatedly encounter this pattern, but often the most important battles are not fought in the courts.

THE RECEDING DARWIN

The Scopes trial ended with a verdict for the prosecution; Darrow had requested this verdict, intent on appealing the conviction to a higher court, he hoped en route to the Supreme Court. Scopes was fined $100, paid by Mencken's paper.

This plan derailed when the Tennessee high court overturned Scopes's conviction on a technicality. No conviction, no appeal. Darrow and the defense team were outraged. Tennessee leadership relaxed, hoping that the Northern papers would cease their barrage of cartoons and articles lampooning their poor state.

Meanwhile, the anti-evolution law in Tennessee remained on the books, joined by a few more passed in other states. Scopes was off the hook, but evolution was not. Concern about grassroots, bottom-of-the-pyramid opposition to evolution motivated textbook publishers to downplay and even remove Darwin from their pages in order to sell more books. This textbook evolution can be seen in the various editions of Truman Moon's *Biology for Beginners*. From 1921 to 1963 this text went through a series of revisions, and each time coverage of Darwin and evolution was reduced. Initially the text had a frontispiece with a picture of Darwin and a meaningful discussion of evolution. Three chapters were removed to accommodate a 1925 Texas anti-evolution law.[9] The 1926 edition dropped Darwin's picture and reduced discussion of evolution, calling it "development." A volume in the 1930s completely removed discussion of human evolution. By the 1950s the word itself had been excised. Each change made the book more popular, and eventually it was the dominant textbook for high-school biology.[10]

In striking contrast to its steady erosion in textbooks, the importance of evolution to the field of biology steadily increased. Within a few decades it was the central organizing principle of the entire discipline. The synthesis of classical Darwinism with the new field of genetics was so compelling that scientific opposition to evolution all but disappeared. A mid-century anniversary essay declared simply: "Biologists one hundred years after Darwin take the *fact* of evolution for granted, as a necessary basis for interpreting the phenomena of life."[11] The tensions that brought Bryan and Darrow to Dayton, however, continued undiminished. They simmered steadily until the 1960s, when they again boiled over, this time in Arkansas.

EQUAL TIME

Separation of church and state is an endless negotiation in America. In 1963 the courts ruled, in a case that would influence the handling of evolution in the public schools, that Bible reading and the Lord's Prayer were not appropriate for public schools. The majority opinion in this case, *Abington School District v. Schempp,* emphasized the importance of a balance in which the public schools would neither advance nor inhibit religion. Critics charged that prohibiting prayer and Bible reading was hostile to religion, serving to establish a "religion of secularism." The court responded that the schools must do nothing to favor "those who believe in no religion over those who do believe."[12]

Evolution returned, invigorated, to the public schools in the 1960s. Cold War competition and a space race with Russia raised concerns about the general weakness of science education in America. Curricular overhaul produced new textbooks across the board. The biology texts were often written in ivory towers by northern academics blissfully unaware that most Americans remained opposed to evolution. Darwin's controversial theory moved onto center stage and became the heart of the entire biology curriculum.

Anti-evolutionists applauded Schempp's demand for neutrality, charging that a curriculum containing only the secular story of origins was far from neutral in that it promoted the "religion" of secularism over traditional Christianity. They demanded "equal time." If a theory of origins *hostile* to Christianity was taught it must be balanced by a *congenial* theory of origins. This was an interesting variation on Bryan's demand at the Scopes trial. Where Bryan wanted balance by teaching *no* theories of origins—a "balance" essentially achieved when textbook publishers all but eliminated coverage of evolution—the new creationist strategy was a balance achieved by equal time for both positions.

Meanwhile, anti-evolution laws remained on the books, remnants of Bryan's populist rampage. The emasculated coverage of evolution made these laws moot, for the most part, but the curricular reform of the 1960s gave birth to textbooks filled with evolutionary biology. Clouds began to appear in the form of concerns that "equal time" needed to advance from an interpretation of the law to mandated practice. The gathering storm grew steadily as evolution took up residence in textbooks, even while the states purchasing those books had legal, if largely ignored, bans on teaching the theory. The past and the present grew increasingly at odds.

Things came to a head in 1968 in a trial eerily reminiscent of the Scopes trial. A young biology teacher, Susan Epperson, challenged Arkansas's law against teaching evolution, claiming, among other things, that it violated her freedom of speech. The years that separated the trials, however, had witnessed dramatic changes in America and things looked very different now.

The Scopes judge considered the issue so simple that he refused to even hear expert witnesses. There was no need for "clarification." In the four decades since Scopes, science had moved steadily forward, while anti-evolutionary sentiment ran in place on a treadmill powered by nineteenth-century arguments. The advances in science displaced the perception that evolution was a speculative theory on the margins of biology. Evolution was now what educated people believed and, of course, children should learn it. Laws against teaching evolution were like the archaic laws still on the books about not leading animals onto the interstate—although there had been no occasion to repeal these legal fossils, their relevance had certainly diminished over the years.

The judge hearing the Epperson case thought the Arkansas statute was ridiculous, reasoning that if evolution was in the biology text, then it should be taught. He scheduled the trial on April Fool's Day to make his point and gave the state just one day to queue up their expert witnesses to make the case against evolution.[13] The exact legal issue was complex, nonetheless. The Epperson lawyers argued the case on the basis of First Amendment freedom: teachers have a constitutionally guaranteed freedom of speech and should not be legally prevented from teaching whatever topics they deem appropriate. This interpretation, of course, is really quite unworkable, at least in theory, for it sets no limits on what a teacher could bring to the classroom. Under this broad freedom astrology, psychic healing, channeling, and alien abductions would all be permitted in public-school classrooms. And, of course, it would be permissible to teach creation science. The defense warned that schools could now be forced to make room for "the haranguing of every soapbox orator with a crackpot theory."[14] Epperson's attorneys responded by asking only that Epperson be allowed to teach what was in the textbook, essentially making their case now identical to that of Scopes. And, just as in Dayton, attempts to engage the truth or falsity of evolution were consistently derailed.

The judge rejected the Arkansas statute as unconstitutional. Epperson and her fellow Arkansas teachers were now free to teach what was in the textbooks, whether or not it offended the religious sensibilities of their students. They were also free to teach creation science, tellingly absent from the textbooks.

ON TO THE ARKANSAS SUPREME COURT

The Arkansas Supreme Court consisted of elected officials in a state where most voters opposed evolution. Biologists from Arkansas did not write Epperson's textbooks, and the Epperson ruling did not sit well with many locals. So perhaps it is not surprising that the court reversed the Epperson ruling a year later, with little reason given, and restored the Scopes interpretation of anti-evolutionary statutes: states should control the public-school curriculum. The obvious motivation was a desire to pass an unpopular buck to the Supreme Court, an act of judicial cowardice that did not go unnoticed by the Supreme Court justices. Even Epperson's concern that she would get in trouble for teaching the evolution in her textbook was viewed with skepticism. Justice Black, intuiting that Epperson was just play-acting in an updated John Scopes role, wrote:

> Now, nearly 40 years after the law has slumbered on the books as though dead, a teacher [Epperson] alleging fear that the State might arouse from its lethargy and try to punish her has asked for a declaratory judgment holding the law unconstitutional.[15]

Fully aware it was playing a game with Arkansas legislators, the Supreme Court struck down the 1928 Arkansas statute in November 1968. Two years later, Mississippi's state Supreme Court struck down its law against teaching evolution. It was the last one standing, and with its departure the legal legacy of Bryan's anti-evolution campaign was finally dismantled. There remained, however, a general enthusiasm in most of the country for the teaching of creationism and an even more widespread conviction that high schools should teach "both sides," giving equal time to both creation and evolution. Anti-evolutionary forces were already at work to create this awkward balance.

UNEASY TRUCE

The 1970s saw great discussion of the *balance* between the teaching of creation and evolution. Most Americans thought creation belonged in the public schools, at least as an option. The scientific community, however, insisted that evolution was the only real science of biological origins. Creationists, searching eagerly for some legal doorway into the public schools, became enthusiastic champions of exposing high-school students to different ideas and then letting them choose. Apparently, they believed that bored sixteen-year-olds, obsessed with dating and their complexions, were better positioned to evaluate theories of origins than the scientific community.

In the early 1970s, the Christian publisher Zondervan produced a polished creationist textbook, *Biology: A Search for Order in Complexity*. Although the text appeared on educational radar screens, it was continually mired in controversy and never became the standard alternative text that its champions hoped. The legal climate was such that this unique text, in principle, could have been widely used. Carefully avoiding mention of biblical ideas, *Biology: A Search for Order in Complexity* was approved for adoption by many state school districts. But when one local district adopted *only* this text, implying that its students would have no evolutionary text at all, the ACLU predictably took the matter to court. Once again the "pyramid problem" worked against the creationists. Although it was approved by a state commission and adopted by a local school board, a judge rejected the text because it "advanced particular religious preferences and entangled the state with religion."[16]

This critique proved to be the Achilles' heel of creationism and its successor, intelligent design. The pattern would repeat. Textbooks, theories, ideas, and even individual scholars that impressed grassroots conservatives would vaporize into irrelevance when confronted with a more sophisticated audience. It would be an enduring challenge to the anti-evolutionary movement—a challenge that would intimidate most, but not all, of the champions of creationism.

One such champion was Yale law student Wendell Bird. In January 1978, Bird published an award-winning article in the *Yale Law Journal* outlining a strategy for getting creationism into America's public schools. Rather than relying on the simple absence of prohibitions to create space for creationism, Bird developed an argument for the mandatory inclusion of both creation and evolution.

Bird dusted off Bryan's old concern that teaching evolution violated many students' religious faith. Government-supported teaching of evolution was an unconstitutional interference in students' religious freedom. Students from traditions that read Genesis literally, in keeping with precedents granted to groups like the Amish and Jehovah's Witnesses, had a right to opt out of instruction incompatible with their faith. But, Bird argued sensibly, opting out of high-school biology was undesirable, and schools should "neutralize" this consequence by teaching creationism alongside evolution. Furthermore, creationism could be recast as pure science, so the "balanced treatment" really amounted to nothing more than a fuller presentation of scientific ideas about origins.

Overnight, Bird became a celebrity in the creationist cause. He was an intellectual heavyweight from the top of the pyramid, taking the torch from Bryan and supported by the same grassroots populism that energized the Great Commoner. Bird joined Henry Morris at the Institute for Creation Research (ICR) and served for a while as its staff attorney. ICR is a multi-pronged, fundamentalist center of anti-evolution, working on many fronts to reverse the "harmful consequences of evolutionary thinking on families and society (abortion, promiscuity, drug abuse, homosexuality, and many others)."[17]

At ICR, Bird drafted a "resolution" promoting a balanced treatment of creation and evolution. The resolution, designed to help interested citizens frame legislation promoting creationism in the public schools, treated both theories as exclusively scientific. Statements based on the resolution circulated broadly, and by 1981 variants had appeared in two dozen state legislatures.[18] The Arkansas Senate passed its version of Bird's resolution by a vote of 22 to 2.[19] The date was Friday, March 13, the same day that the Dayton conspirators in Robinson's drugstore, forty-eight years earlier, had hatched the scheme that led to the Scopes trial.

The ever watchful ACLU took notice and began to move, even as Louisiana passed a similar resolution. It was the beginning of the end of creationism in America's public schools.

THE MIGHTY ACT OF ARKANSAS

Arkansas Act 590 called for a "balanced treatment of creation science and evolution science," defined as follows. *Creation science* means the scientific evidences for creation and inferences from those scientific evidences. Creation science includes the scientific evidences and related inferences that indicate:

1 The sudden creation of the universe, energy, and life from nothing;
2 The insufficiency of mutation and natural selection in bringing about development of all living kinds from a single organism;

3 Changes only within fixed limits of originally created kinds of plants and animals;

4 Separate ancestry for man and apes;

5 An explanation of the earth's geology by catastrophism, including the occurrence of a worldwide flood; and

6 A relatively recent inception of the earth and living kinds.

Evolution science means the scientific evidences for evolution and inferences from those scientific evidences. Evolution science includes the scientific evidences and related inferences that indicate:

1 The emergence by naturalistic processes of the universe from disordered matter and emergence of life from nonlife;

2 The sufficiency of mutation and natural selection in bringing about development of present living kinds from simple earlier kinds;

3 The emergence by mutation and natural selection of present living kinds from simple earlier kinds;

4 The emergence of man from a common ancestor with apes;

5 An explanation of the earth's geology and the evolutionary sequence by uniformitarianism; and

6 An inception several billion years ago of the earth and somewhat later of life.[20]

The Arkansas trial did what Darrow had failed to do at Dayton when his expert witnesses were muted. Arkansas put the creationist *ideas* on trial, exploring whether they were adequately secular for the public schools and scientifically plausible.

As theater the Arkansas trial floundered in the long shadow of Scopes. There was no Bryan-Darrow confrontation, no poor schoolteacher in the dock, no hyperbolic Mencken dressing up the tale for the daily papers, no jury of farmers, and no interesting "North versus South" back story. Reporters called the trial "Scopes II" and made so many references to the original Scopes trial that one journalist commented: "If the readers learned anything, we may assume it is the details of the Scopes Trial."[21]

As an intellectual contest, though, Scopes II made its namesake look like a cartoon. Heavyweights from both sides provided hours of expert testimony, and this time the focus was on the scientific, philosophical, and religious character of creationism and evolution. This case, unlike the charade in Dayton, would not be decided on a technicality.

The intellectual shallowness of the creationist position had pundits on both sides predicting defeat for the statute before the trial even began. Pat Robertson, to whom God apparently speaks directly, accused the Arkansas attorney general, Steven Clark, of being "crooked" and "biased." Robertson suggested that the ACLU had targeted Arkansas because Clark was secretly on its side. The late Jerry Falwell, Robertson's fellow prophet from Virginia, made similar charges.[22] Tellingly, no evidence emerged to support these charges. Clark was simply outgunned by the high-powered legal help that the ACLU brought in from New York.

The expert witnesses fared no better. The stark contrast that results from lining them up side by side makes its own argument for why creationism does not belong in the public schools. To establish whether creation science is religion or science, each side called on testimonies from experts in theology and philosophy. University of Chicago professor of theology Langdon Gilkey testified for the plaintiffs and came across as clever and articulate. At one point he testified that from the perspective of Christian theology, Act 590 contained an egregious heresy. Spectators in the courtroom audibly gasped. In portraying the act as completely secular, Gilkey argued, its supporters had been forced to clarify that the "creator" presupposed in the act and responsible for the "sudden creation of the universe, energy, and life from nothing" was not necessarily "God" as understood in the Judeo-Christian tradition. This, exulted Gilkey, "was precisely the early heresy of Marcion and the Gnostics (about 150 to 200 A.D.), who said that there were in fact two Gods, one a blind, cruel but powerful God of creation ... and the other a good loving God of redemption ... and thus the creator God was *not* the same as the redeemer God."[23]

Gilkey's point had great drama, and he certainly enjoyed its retelling. However, from a legal perspective, his point is irrelevant. The defense was arguing that Act 590 was *not* religious; whether it contained a formal religious heresy should have been of no consequence. But one has to wonder how it was that the fundamentalist Christians who had shepherded this act could have missed this point.

Norman Geisler, then professor of systematic theology at Dallas Theological Seminary, the intellectual heart of American fundamentalism, was

Gilkey's counterpart for the defense. Geisler, a major fundamentalist scholar, has written over fifty books and hundreds of articles. Unfortunately, he embarrassed himself at Arkansas, becoming the brunt of countless jokes in the media about the state's "expert" theological witness. Geisler, like most fundamentalists, believes in a literal devil, the biblical Satan, and in demons. In his deposition for the trial he stated that he had "known personally at least twelve persons who were clearly possessed by the devil." Further evidence for the reality of Satan at work in the world came from UFOs, which Geisler said represented "the Devil's major, in fact, final attack on the earth." And then, to ensure that he completely buried himself, he claimed to know that UFOs were real because he "read it in the *Reader's Digest.*"[24] When he repeated these remarks on the witness stand, the courtroom audience literally laughed out loud; over the next few days newspaper accounts of the trial presented Geisler's remarks as if they were his entire testimony. Lost were his more credible comments about the religious character of evolution and the role of assumptions in science.

The expert witnesses on science were equally mismatched. The plaintiffs had Francisco Ayala, one of the world leading geneticists and a former Dominican priest; Harvard's Stephen Jay Gould, a leading paleontologist and America's greatest science essayist; G. Brent Dalrymple, who had been on the NASA team that investigated moon rocks; Michael Ruse, a leading philosopher of biology; among others. The defense counterparts paled in comparison, a collection of relatively unknown scholars from obscure institutions. There was one major exception, an astronomer named Chandra Wickramasinghe, notorious for proposing that life on earth was "seeded" from outer space rather than developing here:

> The facts as we have them show clearly that life on Earth is derived from what appears to be an all-pervasive galaxy-wide living system. Terrestrial life had its origins in the gas and dust clouds of space, which later became incorporated in and amplified within comets. Life was derived from and continues to be driven by sources outside the Earth, in direct contradiction to the Darwinian theory that everybody is supposed to believe.[25]

Wickramasinghe believed that aspects of evolution were highly implausible. Convinced he had been misled about the credibility of evolution, he supported the general idea that alternative explanations belonged in the public schools. He was, however, hardly in the camp of the creationists and completely rejected almost all the tenets of Act 590. Why the defense called him is curious. Perhaps, in its conviction that there were only two possible positions on origins, the defense inferred that anyone not firmly in the Darwinian camp would necessarily be in the creationist camp. The "two models" approach to origins did, in fact, presume that there were just two models and no others.

The defense attorneys apparently embraced the general and damning creationist confusion that there are only two models of origins. This peculiar oversimplification assumes the two models are in such contradiction that evidence *against* one of them counts as evidence *for* the other. Ayala found it necessary to "educate" attorney David Williams on this elementary point of logic.

"My dear young man," said Ayala, looking at Williams with what Gilkey described as "evident pity," "negative criticisms of evolutionary theory, even if they carried some weight, are utterly irrelevant to the question of the validity or legitimacy of creation science. Sure you realize that *not* being Mr. Williams in no way entails *being* Mr. Ayala!"[26]

With that, Mr. Williams neatly folded his legal tail between his legs and slunk back to his table. "No more questions, your honor."

The creation scientists who were called[27] made it clear that their primary allegiance was to the Bible, not to science. And, although they were confident the two could not conflict, they would set aside scientific findings that disagreed with a literal reading of the Bible.

JUDGE OVERTON'S DECISION

The presiding judge, William Overton, ruled against Arkansas Act 590, finding it religious rather than scientific and likely to harm those students whose education would have been affected by it:

> Implementation of Act 590 will have serious and untoward consequences for students, particularly those planning to attend college. Evolution is the cornerstone of modern biology, and many courses in public schools contain subject matter relating to such varied topics as the age of the earth, geology, and relationships among living things. Any student who is deprived of instruction as to the prevailing scientific thought on these topics will be denied

a significant part of science education. Such a deprivation through the high school level would undoubtedly have an impact upon the quality of education in the State's colleges and universities, especially including the pre-professional and professional programs in the health sciences.[28]

Overton's decision attracted much attention and was widely reprinted by various publications, including America's leading scientific journal, *Science*.[29] The creationists, of course, were disappointed and disagreed. Duane Gish charged that Overton's decision essentially established that "secular humanism will now be our official state-sanctioned religion."[30] Geisler charged that the decision would have "devastating consequences for the pursuit of truth in the public schools."[31]

Philosophers had mixed reactions. Conservative philosopher J. P. Moreland devoted much of his book *Christianity and the Nature of Science* to a critique of Overton's decision.[32] Secular philosopher Larry Laudan, while agreeing with Overton's conclusion, assaulted his reasoning with words like "specious," "egregious," "dubious," "opaque," "woeful," and "silly."[33] But nobody pays much attention to the hair-splitting commentaries of philosophers, and Overton's decision is still widely quoted with approval.

LAST MAN STANDING

By the time the media finished poking fun at poor Norman Geisler for his views on UFOs, statutes mandating equal time for creationism were almost extinct, save for one lone survivor, hiding from the ACLU in the bayous of Louisiana. Called the "Balanced Treatment Act," it was a clone of the statute that had been defeated in Arkansas. It would be creationism's last attempt to sneak into the public schools.

The act stated that schools "shall give balanced treatment to creation science and to evolution science." This balance applied to lectures, texts, and library materials and required the identification of appropriate creation-science materials. The act further specified: "When creation or evolution is taught, each shall be taught as a theory, rather than as proven scientific fact." Creation science was inadequately described as "the scientific evidences for creation and inferences from those scientific evidences."[34] Evolution science had a correspondingly circular definition.

The Louisiana act, unlike its Arkansas sibling, did not define creation in ways that tied it to the Bible. By leaving creation vague and undefined, the religious connection had to be inferred, which everyone in Louisiana, except Wendell Bird apparently, found easy to do. Bird would later argue the case before the Supreme Court, finally getting a "secularized" creation science the hearing he thought it deserved, but first he had to navigate a complex legal maze that would have deterred anyone with less than total dedication.

A COMEDY OF DETOURS

On behalf of Louisiana senator Bill Keith, Bird launched a campaign to force Louisiana schools to implement the Balanced Treatment Act. Implementation was almost nonexistent, probably because it was impossible to find suitable educational materials. The ACLU, still flushed from its victory in Arkansas, responded the next day. On behalf of Daniel Aguillard, it attacked the Louisiana act as unconstitutional, arguing that the mandated creation science was still the Judeo-Christian creation story despite the careful secularization. Furthermore, the ACLU noted that the act's history revealed consistent support from fundamentalist Christians, further evidence of its religious character. Bird would later challenge this, arguing that a position is not automatically religious just because it has religious roots or religious supporters.

Bird was appointed special assistant attorney general for the state of Louisiana and became the point man in the upcoming trials. He hoped to finally square off against the ACLU on the constitutionality of teaching a secularized creation science. He demanded a trial, but was soon frustrated, as both trials began to mutate, evolve, and stagger toward extinction.

The court looking at the Keith motion to implement the Balanced Treatment Act decided that no constitutional issue was at stake and the Louisiana courts could decide this matter internally. The Keith motion would not lead to a constitutional review of creation science.

The Aguillard court ruled that the Louisiana legislature could not tell the state education board how to run the schools. It insisted on an unusual, and perhaps contrived, autonomy for the state education board. The judge ruled that the education board should decide how, and if, the Balanced Treatment Act would be implemented. A frustrated Bird appealed this decision and won. The trial, which had been a mirage, began to take shape, although not in Louisiana.

Bird built much of his case on five affidavits he had prepared for the elusive earlier trial. Two scientists, a philosopher, a theologian, and a school administrator had prepared briefs arguing for the legality of teaching a religiously neutered version of creation science. Bird intended to present these expert opinions, hoping the judge would find them convincing. The Aguillard judge, however, was not impressed. He ruled that the Balanced Treatment Act was religious despite the claims in the affidavits. Even though the ACLU had not followed standard practice by bringing countering affidavits, the judge was not persuaded to give Bird his trial on the merits of creation science.

Bird, of course, was far more than a lawyer arguing one side of a case. He believed passionately that creation science was a true account of origins, supported by overwhelming scientific evidence. He was equally convinced that evolution rested on flimsy evidence and had a checkered history of false claims and disturbing applications. Like so many creationists, Bird was alarmed that America was not teaching the truth about origins to its children. If God created the earth ten thousand years ago, this event was a *scientific* fact, regardless of what the Bible or any religion might say on the matter.

Bird convinced Louisiana to appeal the decision, and in 1985 the appeals court agreed to hear the case. Bird pointed to the five affidavits that the ACLU had not challenged with countering affidavits. Was this not, inquired Bird, because these affidavits were so compelling there were no effective challenges? Technically, affidavits presented by one side should be accepted as true unless countered by opposing affidavits. The earlier decision declaring the Balanced Treatment Act religious should therefore be overruled.

Bird lost again. The judges were not convinced by the affidavits and ruled the act was religious, with no secular purpose. Bowed but unbeaten, Bird challenged their decision. Invoking a technicality, he demanded a ruling from all fifteen of the judges on the court; if the majority agreed with the full court press, then they would all have to participate. But it was not to be. Eight of fifteen judges disagreed, and Bird lost again, by one vote.

The seven dissenting judges, however, indicated some support for creation science, agreeing it was constitutionally feasible to teach creation science in the public schools. After years of frustration, Bird finally received encouragement that high-level legal opinion might actually come down on the side of creation science. His sights began to set on Washington, D.C., where nine of the nation's leading judges could resolve, once and for all, the constitutionality of teaching creation science in America's public schools.

ON TO THE U.S. SUPREME COURT

Bird's petition for a Supreme Court review referenced the five unchallenged affidavits ignored by the Louisiana judges. The court agreed to hear the case on December 10, 1987.

Despite the years of preparation involved in getting this case to the U.S. Supreme Court, when the legal ball game finally got started, it was Scopes and Arkansas all over again. When the best arguments from both sides were lined up, there was simply no competition. Bird's affidavits and other briefs filed by creationist groups looked like high-school projects alongside the opposing arguments assembled by the ACLU. Bird's affidavits included two by scientists. One was Dean Kenyon, a minor, although competent, scientific figure from San Francisco State University whose reputation was derived almost entirely from his support for creationism. Kenyon argued:

> It is also my conclusion that balanced presentation of creation-science and evolution is educationally valuable, and in fact is more educationally valuable than indoctrination in just the viewpoint of evolution. Presentation of alternate scientific explanations has educational benefit, and balanced presentation of creation-science and evolution does exactly that. Creation-science can indeed be taught in the classroom in a strictly scientific sense, and a textbook can present creation-science in a strictly scientific sense, either as a supplement or as a part of a balanced presentation text.[35]

The other scientist was W. Scott Morrow, who taught chemistry at a small religious college. Like Kenyon, Morrow had little reputation beyond what he acquired as an advocate of creationism. In the Arkansas trial Morrow noted that it would be interesting to teach flat-earth theory in the public schools, so it was not obvious what criteria he was using to endorse school curricula.[36] Morrow also called himself an agnostic and an evolutionist.

Affidavits from prestigious scientific groups in support of the ACLU challenged those of Bird. The National Academy of Sciences wrote one, as did seventeen state academies of science. One ACLU brief was signed by seventy-two Nobel laureates in science. Under the bright lights of this Supreme Court case, this head-to-head battle once again looked like an embarrassing contest.

Bird's presentation didn't fare much better on the religious side. Two of his venerable affidavits were from minor scholars of religion: Terry L. Miethe, then at Liberty University, fundamentalism's most powerful bastion of higher education but without respect in the secular world; and William G. Most, of Loras College, a Catholic school. Miethe and Most both argued that, although they did not themselves accept creation science, it was indeed possible to teach it in fully secular manner.

The Supreme Court case laid out the question exactly as Bird had long wanted. Bird, clutching his briefs and affidavits, argued that creation science met the criteria for science at least as well as evolution. Both dealt with a murky history that was hard to interpret. Both invoked processes, such as the origin of life or the appearance of the universe, that are not presently occurring and thus cannot be studied directly. Both were complex, with elaborate logical and empirical structures.

Addressing religious concerns, Bird argued that creation science could be secularized for the public schools. This, in fact, was precisely the difference between "creation" and "creation science." If the universe originated ten thousand years ago, as creation scientists claim the data indicates, presenting evidence for this in the public schools is not inherently religious. Neither is it religious to note that the fossil record contains fewer intermediate forms than evolutionists would like, and that most new species appear in that record suddenly. This was, in fact, exactly what America's best-known evolutionist, the late Stephen Jay Gould of Harvard, had long been calling paleontology's "dirty little secret."[37] There is nothing religious about noting the absence of any generally accepted theory explaining the origin of life. Scientific claims don't become religious just because religious people like them.

Bird waxed eloquent. There are two different, incompatible models for origins. One suggests that everything evolved slowly from simpler forms; the other asserts that everything appeared suddenly. Why not teach both models in high-school biology, exposing students to a broader range of ideas? Shouldn't we encourage critical thinking on the part of our students? Shouldn't we allow them to weigh the evidence and make up their own minds?

Creation *science,* argued Bird, was not religious. The lower courts in Louisiana had pulled that idea "out of thin air,"[38] and the Supreme Court justices would surely know better. Requiring that evolution be balanced with creation science was Louisiana's way of promoting fairness, giving all sides a place on the chalkboard, and ensuring academic freedom for teachers. The goal here was simply providing the best possible curriculum for the students.

Shortly after the trial, in a massive two-volume survey of the topic, Bird wrote:

> The issue is *not* which explanation of origins is correct, but whether any is so compellingly established and universally accepted that it ought to be taught to the exclusion of other scientific explanations. Because no theory is so unquestionably true, all scientific views should be taught to protect the students' right to receive scientific information and the teachers' right to academic freedom by offering the "whole scientific truth."[39]

BYE-BYE BIRDIE

Mounting a more effective defense for creation science than what Bird presented to the Supreme Court on that Wednesday in Washington, D.C., would be hard. The contours of the legal arguments that had been circulating for the past decade had been outlined in his influential paper at Yale. He understood as well as anyone exactly how the First Amendment applied to this issue. Bird had worked closely with the world's leading creationist organization, the Institute for Creation Research, and knew all the key players. He was familiar with the scandal of the Scopes trial and had watched in horror as the creationists humiliated themselves in Arkansas. And he had nurtured the Louisiana case from infancy to its full maturity before the highest court in the country, where a victory could have transformed the teaching of origins across the country.

Nevertheless, Bird lost.

Bird's arguments were compelling and in a perfect world with no history might have carried the day. The Supreme Court justices certainly did not sneer and laugh, as had their counterparts in Arkansas. William Brennan wrote the majority opinion. Speaking for seven of the nine justices, he rejected Bird's argument that the Louisiana statue was truly secular. Brennan was known for preferring a "high" wall of separation between the state and religion, in part because he believed religion was "too important to be co-opted by the state."[40]

The two justices most interested in religion (and probably most personally religious) disagreed with the majority opinion. William Rehnquist, an active Lutheran, joined Antonin Scalia, a conservative Catholic whose son Paul was a priest, in a dissenting opinion. Scalia accused his fellow justices of being blinkered by the Scopes legacy and argued, echoing Bryan, that the people of Louisiana were entitled "to have whatever evidence there may be against evolution presented in their schools."[41]

Bird's argument, in the final analysis and perhaps even independently of his own understanding of the law, was a Trojan horse. As noble as it might seem to "balance" education, the reality was that creation science was nothing but a tiny intellectual backwater championed by a handful of minor fundamentalist scientists.[42] If every tiny opposing viewpoint received the equal time that Louisiana wanted for creation science, the public schools would be opening their doors to astrology, Holocaust denial, alien visitation, and countless other preposterous topics.

The long history of creation science revealed its thoroughly religious pedigree. Unlike evolution, whose adherents include many Christians of almost every variety as well as agnostics and outspoken atheists, creation-science adherents were almost exclusively fundamentalist Christians. This was not a coincidence and support for creationism has consistently been driven by a particular enthusiasm for biblical literalism, not scientific data. Furthermore, creationism has scant support among educated theologians and biblical scholars.

CREATIONISM EVOLVES

One of my favorite Monty Python skits involves a strange conversation in which a man with many pets named Eric tries to buy a license from a shopkeeper for his fish, Eric. At one point the pet owner states that he has a license for his cat, Eric. The shopkeeper responds that there is no such thing as a cat license. Defending his claim that there is, the eccentric pet owner triumphantly presents written documentation as evidence. Upon examining the documentation, the shopkeeper responds, "This isn't a *cat* license. It's a *dog* license with the word 'dog' crossed out and 'cat' written in, in crayon." The next major trial involving creationism in America's public schools gave new meaning to this absurd skit.

The Supreme Court's ruling in *Edwards v. Aguillard* derailed creationism's ride into the public schools by legal mandate. Creation science could not be decoupled from the Bible to the court's satisfaction, even by clever legal strategists like Bird. However, rather than going extinct, creationism evolved rapidly and before long reappeared in a dramatic new guise known as intelligent design, or "ID" for short.[43] ID's champion was Phillip Johnson, a legal scholar then at Berkeley's prestigious Boalt Hall law school.

Johnson recognized that the most theologically important issue in the origins controversy was not creation versus evolution, but the exclusive reliance of the natural sciences on purely naturalistic explanations. Science had come to the point where, by definition, *nothing* could ever be explained by reference to God. This naturalism, in Johnson's mind, was equivalent to atheism. In a hyperbolic generalization of the sort that came to characterize his polemical style, he rearticulated the traditional anti-evolutionary argument: (1) the institutions of modern society are based on science; (2) science is based on atheism; and (3) a society with atheistic foundations will quickly go to hell in a handbasket, just as Western civilization is presently doing.

Johnson developed a legal strategy for continuing the fight against evolution not unlike what Bird had done a decade earlier. But he used a new tactic, making no reference to creation or a creator and completely avoiding anything even resembling the biblical story of origins. Johnson simply called attention to complex phenomena in nature that posed problems for evolutionary explanations. He argued that these phenomena could be explained only by invoking an outside intelligence. "Intelligent design," as this strategy became known, was defined as the theory that "various forms of life began abruptly through an intelligent agency, with their distinctive features already intact."[44]

Johnson and the enthusiasts who jumped on his bandwagon insisted that by not specifying the identity of the "intelligent agency" they had at last fully secularized their "theory." The movement gathered steam throughout the 1990s and attracted scholars like biochemist Michael Behe, mathematician-philosopher William Dembski, biologist Jonathan Wells, philosopher Stephen Meyer, and biologist Dean Kenyon. Adopting the politically expedient philosophy that "the enemy of my enemy is my friend," the ID tent grew very large, welcoming any and all opponents of evolution. ID embraced creationists of all stripes, young-earth and otherwise, and even welcomed the occasional non-Christian anti-evolutionist. Supporters of ID shared one—and sometimes only one—central belief: there is design in nature that evolution cannot explain. Disagreements like whether the earth is thousands or billions of years old were set aside as unnecessarily divisive.

Johnson's leadership, a handful of surprisingly popular books, and ID's big-tent strategy combined to grow an impressive, popular movement that fanned the waning flames of anti-evolutionary activism. Countless school boards, populated as they were by ordinary citizens, passed initiatives weakening the teaching of evolution, convinced that serious objections to biology's central concept were being established. Some of these initiatives were cancelled democratically, by simply voting out members of the school boards responsible. Other initiatives went to court. The intelligent design movement got its fifteen minutes of fame in late 2005 in Dover, Pennsylvania.

CREATIONISM IN DESIGNER CLOTHING

The Dover trial was something of a replay of the case Bird lost before the Supreme Court. The challenge facing Bird in that trial was establishing that creation science was secular and thus *not* the same collection of ideas that had been ruled religious in the *McLean v. Arkansas Board of Education* trial. Likewise the Dover trial turned on the question of whether ID was a secular concept or a repackaging of creationism.

The Dover story hit the news on November 19, 2004, when the local school district issued a press release stating that, come January, teachers would have to read the following statement to students in ninth-grade biology:

The Pennsylvania Academic Standards require students to learn about Darwin's Theory of Evolution and eventually to take a standardized test of which evolution is a part.

Because Darwin's Theory is a theory, it continues to be tested as new evidence is discovered. The Theory is not a fact. Gaps in the Theory exist for which there is no evidence. A theory is defined as a well-tested explanation that unifies a broad range of observations.

Intelligent Design is an explanation of the origin of life that differs from Darwin's view. The reference book, *Of Pandas and People,* is available for students who might be interested in gaining an understanding of what Intelligent Design actually involves.

With respect to any theory, students are encouraged to keep an open mind. The school leaves the discussion of the Origins of Life to individual students and their families. As a Standards-driven district, class instruction focuses upon preparing students to achieve proficiency on Standards-based assessments.

There were two questions on the table in Dover: Is ID something other than creationism? Is *Of Pandas and People* a creationist book?

Reminiscent of Scopes, Dover attracted big legal guns itching for a fight. Much of the energy behind the Dover initiative came from the Thomas More Law Center, the self-proclaimed "Christian Answer to the ACLU." Thomas More aggressively sought confrontations with the ACLU on all of the standard issues such as gay marriage, pornography, public displays of the Ten Commandments, and nativity scenes. Representatives from Thomas More had been encouraging school districts across the country to teach ID and authorize the *Pandas* book as a supplemental biology text. Knowing that the ACLU would eventually challenge such decisions, Thomas More promised to defend, for free, any school that got sued.

Sure enough, on December 14, 2004, the ACLU filed suit against the Dover school district on behalf of some parents with school-age children. A call went out for a big law firm to provide pro bono legal help. Eric Rothschild, a partner in a major Philadelphia firm, quickly volunteered, saying "I've been waiting for this for fifteen years."[45]

Because of the role played by precedent in the American legal system, the Dover trial was not really about ID per se, but rather turned on the somewhat simpler question of whether ID was, as Leonard Krishtalka, who directs the Natural History Museum at the University of Kansas, put it, "creationism in a cheap tuxedo."[46] The courts had already established that creationism was religious and could not be taught in the public schools. If it could be established that ID was a form of creationism, then precedent mandated that it had no place in the public schools.

Dover was a disaster for ID. The ACLU established in a variety of ways that ID was indeed a dressed-up version of creationism. Adding salt to the already severe wounds, it emerged that key ID people—deeply religious people—in the trial were actually lying and knowingly misrepresenting their case.

The 1989 *Pandas* book, at the center of the Dover controversy, provides an excellent example of this deception. ID pundits claimed that the text was the "first intelligent design textbook,"[47] and *not* a creationist text. Unfortunately, there existed damning early drafts of the book from before the Aguillard ruling that creation science could not be taught in the public schools. These earlier versions revealed that the original plan was for the book to be a creationist text. A 1983 draft was even titled *Creation Biology;* a 1986 draft was titled *Biology and Creation* and contained the following definition: "Creation means that the various forms of life began abruptly through the agency of an intelligent creator with their distinctive features already intact. Fish with fins and scales, birds with feathers, beaks, and wings, etc."[48]

In a Monty Pythonesque editorial move, the post-Aguillard edition of the book replaced "creation" with "intelligent design" and left the rest of the definition virtually unchanged: "Intelligent design means that various forms of life began abruptly through an intelligent agency, with their distinctive features already intact—fish with fins and scales, birds with feathers, beaks, wings, etc."[49]

The defense attorneys then argued that *Pandas* was *not* a creationist book, but rather was about something entirely different. To say the least, this was exceedingly disingenuous. They certainly knew better. If the changes had been made in crayon, Monty Python could have sued them for stealing their skit.

To make matters worse, the writers who had produced *Pandas* had strong connections to creation science. Dean Kenyon, the lead author, had written the foreword to *What Is Creation Science?* in which he proposed "that all students of the sciences ... should be taught the major arguments of both the creation and evolutionary views."[50] The second author, Percival Davis, had coauthored *A Case for Creation.*[51] Journalist Nancy Pearcey, who made major contributions to *Pandas,* was a young-earth creationist and editor of the *Bible Science Newsletter,* where portions of *Pandas* had been excerpted.

Witnesses from the Dover school board testified that there had been considerable support for creationism on the board. The chair of the curriculum committee, William Buckingham, denied that he had supported creationism, but multiple witnesses and stories in local papers all reported that he had been arguing that evolution must be balanced with creation. Although he denied it in his deposition, Buckingham was quite passionate about this, convinced he was doing God a favor: "Two thousand years ago, someone died on a cross," he said. "Can't someone take a stand for him?"[52] Though he claimed to have no knowledge of the source of the funds used to purchase and distribute copies of *Pandas,* it turned out he had raised the money himself from his church. Such duplicity plagued the defense to the point that Judge Jones actually got angry and started asking questions. In his decision he made reference to these "flagrant and insulting falsehoods,"[53] noting as "ironic" the contradiction between defendants who "staunchly and proudly touted their religious convictions in public," but then in the trial would "time and again lie to cover their tracks."[54]

The Thomas More lawyers needed to show that the Dover legislation was not motivated by a religious agenda in order to avoid one of the criteria used to label such initiatives unconstitutional. There was, however, no way to hide the local religious enthusiasm for something other than evolution in the schools.

The leaders of the ID movement had also damned themselves on the record. The leading ID think tank, the Seattle-based Discovery Institute, had produced a strategic plan for the widespread promotion of ID. Called the "Wedge Document," the plan stated that the goal of ID was "to replace materialistic explanations with the theistic understanding that nature and human beings are created by God." The ID movement was mainly a populist crusade against evolution, nurtured by the same grass roots that energized Bryan eight decades earlier. ID leaders had all written voluminously for their primary audience, conservative evangelicals. Such writings, usually published by conservative Christian presses, were filled with discussions of how ID supported a biblical worldview, how ID helped prove the existence of God, and how evolution was just atheism in disguise. Now, in a setting where ID had to be secular to survive, its deeply religious character was clearly visible beneath a thin veneer of secular rhetoric.

The defense didn't fare much better on the science side. The Discovery Institute actually got worried that things were going so badly it convinced some of its "fellows" not to participate, lest their reputations go down in the flames they saw being kindled.

Michael Behe testified anyway, as the star witness for the defense. As the author of the best-selling *Darwin's Black Box,* published by the respected and secular Free Press, and one of a small number actively publishing scientists in the ID movement, Behe was something of a celebrity. But his performance on the witness stand was somewhere between ineffective and disastrous. He admitted, for example, that he did not agree with the description of ID in *Pandas,* despite the fact that he was listed as a "critical reviewer" of the book. He admitted being unfamiliar with many major research studies contradicting claims he had made himself in *Darwin's Black Box*. He admitted that neither he nor anyone else had actually developed any "quantitative criteria for determining the degree of complexity,"[55] a critical first step in making ID scientific. Devastatingly, he admitted that changing the definition of science to include ID would also bring astrology into the scientific fold.

The judge quoted extensively from Behe's testimony in his remarks, including the following damning admission: "There are no peer reviewed articles by anyone advocating for intelligent design supported by pertinent experiments or calculations which provide detailed rigorous accounts of how intelligent design of any biological system occurred."[56]

Despite being a respected, competent, and well-published biochemist at a major university, Behe projected a persona in Dover that was of a lone and eccentric outsider with idiosyncratic and occasionally confused notions about science. Even his own department at Lehigh issued a statement of nonsupport for Behe's work on ID. Behe did, of course, have considerable stature within the ID movement itself. ID proponents hoped that at Dover he would appear to be the voice of a growing movement of mainstream scientists dissatisfied with evolution. It didn't happen.

THE DOVER RULING: "BREATHTAKING INANITY"

Judge Jones issued his ruling shortly before Christmas, indicting ID on every front. Any reasonable person would know, he wrote, that the ID strategy was nothing more than a continuation of the failed strategies employed by "earlier forms of creationism." The support for ID in Dover was rooted in local fundamentalist fervor and started not with concern about science in the schools, but concern about the absence of religion. Several board members admitted knowing nothing about ID other than that getting it into the schools would undermine evolution and advance creation. One board member, strongly on the side of the defense, didn't even know what ID stood for, referring to it as "intelligence design."[57] And in the months leading up to the trial the Thomas More law firm was cheering from the sidelines, eager to meet the ACLU in court and confident that that "God was on their side."

The disclaimer the defense wanted read in the classrooms, wrote Jones,

> singled out the theory of evolution for special treatment, misrepresents its status in the scientific community, causes students to doubt its validity without scientific justification, presents students with a religious alternative masquerading as a scientific theory, directs them to consult a creationist text as though it were a science resource, and instructs students to forego scientific inquiry in the public school classroom and instead to seek out religious instruction elsewhere.[58]

As earlier trials had revealed, creationists had no alternative science of their own. *Pandas* was a relabeled creationist text containing little more than a list of things not adequately explained by evolution. It was hopelessly out of date and could only be mistaken for a science text by readers who knew nothing about science. *Pandas* was a religious book with the word *religion* crossed out and the word *science* written in.

Dover was a tragic defeat for ID, the tragedy compounded by the hopeless disorganization of the defense. A naive and uninformed school board, bewitched by a century of anti-Darwinism and cheered on by a zealous law firm, failed to make even its own best case. Completely missing was a new Wendell Bird, who, although he lost at the Supreme Court, had at least made the best possible case for creation science. The judge suggested that the Dover school board's actions constituted "breathtaking inanity" and precipitated a pointless trial that was an "utter waste of monetary and personal resources."[59]

ID could certainly have turned in a more impressive performance. The politically savvy Discovery Institute probably acted wisely in minimizing its involvement, given the debacle it saw developing. Nevertheless, once the damage was done, it tried to contain its impact by publishing aggressive critiques of the judge's decision.[60] These initiatives were clearly disingenuous, given that the institute had refused to cooperate. By refusing to be involved in the trial, the Discovery Institute had forsaken its opportunity to help Judge Jones get the clearest possible picture of ID, or at least their version of the picture.

The Dover decision remained a local ruling, which means that it applies in Dover and nowhere else. But the trial's national publicity, like that of Dayton eight decades earlier, certainly discouraged at least some similar initiatives elsewhere.

THE NEVER ENDING TRIAL

America's creation–evolution trials make for great drama—literally in the case of Scopes, but also on smaller stages with lesser-known protagonists. Each trial offers its own window into the fears and frustrations of ordinary people as they struggle with a science threatening their faith. From William Jennings Bryan in Dayton to William Buckingham in Dover, the trials were never truly about science. Bryan thundered his anti-evolution message across the country because he wanted to protect ordinary Americans from those who "have no other purpose than ridiculing every Christian who believes in the Bible."[61] Buckingham labored to make space for creationism in Dover for similar reasons. Bryan and Buckingham are typical anti-evolutionists. Not scientists, both were enthusiastic Christians concerned about the pernicious effects of evolution steadily eroding traditional American values.

Every one of the trials, and countless smaller episodes that did not make it to trial, had its Bryan or Buckingham. Always in the background were local churches, where people prayed and pastors promoted. The scientists who testified were almost always outsiders to the community in more ways than one; sometimes they were atheists. The creationists were often well-known Christian leaders, "brothers and sisters in Christ." The deliberations inevitably took on an apocalyptic character as the forces of "good" and "evil" locked horns in a conflict that was cosmic, not local. When Clarence Darrow strolled down Main Street in Dayton, two local women called him a "damned infidel."[62] When Cornell sociologist Dorothy Nelkin admitted under oath in the Arkansas trial that she did not believe in a personal God, three spectators in the court fell dramatically to their knees and began to pray audibly for her soul.[63]

Anecdotes like these were enlarged in Mencken's sarcastic reporting, in the enduring presentations of *Inherit the Wind,* and in newspaper accounts of the subsequent trials. Eccentric and extreme players on both sides were juxtaposed as representative, turning the trials into caricatures. Creationists were inevitably portrayed as corny Southern hillbillies, or their Northern equivalent, and freakishly religious; evolutionists were articulate, agnostic, and sometimes antireligious. Pitting them against each other was great journalistic fun.

America's long legal struggle with evolution, however, is anything but a war of religious hillbillies against Ivy League agnostics. It is the ongoing story of a deeply religious nation, with enduring populist and even anti-intellectual sentiments, struggling with an emerging secular science. The trials, when viewed as a whole and separated from the colorful personalities who make them so interesting, reveal a remarkably coherent story.

The Scopes trial culminated Bryan's anti-evolution crusade that had state legislatures across the South outlawing the teaching of evolution. The energy for this crusade, however, did not come from widespread concern that evolution was incompatible with the Bible, although that was certainly a background issue. The energy came from the belief that evolution was the foundation of evil social agendas. In this sense the anti-evolutionary campaign was more like the war on drugs than a war of ideas.

The stakes were smaller the next time evolution went to court, and all that was asked was that creationism be taught alongside evolution. Finally, at Dover, an even more modest proposal went on trial: that evolution be taught with recognition that it was flawed. Each time, evolution, post-Dayton, emerged victorious.

Polls, however, continued to show unwavering support for creationism regardless of what was legally mandated for America's science classrooms. A 2005 survey conducted by *CBS News* revealed that 51 percent of Americans believe that God created humans in present form.[64] The steadily increasing credibility of evolution and its embrace by the educated elite were more than offset by successful grassroots campaigns that maintained the anti-evolutionary fervor whipped up by Bryan decades earlier.

The caricatures of the trials that took up residence in American culture neatly divided the opponents into "science" and "religion" camps. Creationism became the "religious" position, in the eyes of many, and evolution the "scientific" position. But the actual history is much different.

I argued in an earlier chapter that creationism was not, from a religious point of view, particularly important at the beginning of the twentieth century. At Dayton potential religious witnesses sided with Scopes, although their testimony was precluded as irrelevant. The ACLU's strategy included showing that the conflict was not between "religion" and "science," but between a religion that was keeping up with science and one that was not. At Dayton, in Arkansas, at the Supreme Court, in Dover, and on every legal field where creation and evolution met, there were always strong religious voices in support of evolution. Biblical scholars and theologians from all but the most conservative Christian denominations were every bit as opposed to creationism as the scientists from their ivory towers. I have found, for example, after more than two decades as a faculty member at an evangelical college, that the most vigorous opposition to creationism comes from scholars in religion departments rather than in scientific disciplines. As strong as the scientific evidence against creationism has become, the biblical and theological arguments for rejecting it are perhaps even stronger. Expert scholars of religion made this clear in each of the trials.

But Americans have never been eager or even willing to be led by intellectual elites. A simple commonsense argument by someone you trust is worth more than the pompous pronouncements of an entire university of condescending eggheads. America is a nation that loves cowboys, and cowboys don't need experts telling them what to think.

CHAPTER 5

THE EMPEROR'S NEW SCIENCE

S cientists know the moon to be two hundred and forty thousand miles away. How would you react if your neighbor, who was very interested in science, said it was a quarter mile away, closer than the convenience store you can see from your front step? Imagine attending a massive rock concert that broke all attendance records with ten million fans. Your neighbor, who was in attendance, claims there were just ten fans at the concert. Suppose you discussed the age of the earth with your neighbor. In agreement with scientists, you say the earth is five billion years old; your neighbor, however, says that number is a million times too large and the true age of the earth is just over five thousand years. Such extreme disagreements seem laughable and artificial. The last one, however, is a highly animated argument in America as young-earth creationists, a hundred million strong, spar with the scientific community over the age of the earth. Nobody thinks the moon is just above the rooftops, but most people in America have a neighbor who thinks the earth is ten thousand years old.

Creationists disagree with mainstream science on many topics, preferring their own alternative creation science. We hunt in vain, though, to find a more dramatic numerical disagreement on any topic than the one that exists in America today over the age of the earth. If the number of creationists was small, say, comparable to the group claiming abduction by aliens, this would be nothing more than a curious example of human eccentricity. But a 2006 Gallup poll indicated that almost half of all Americans are bona fide creationists, agreeing that "God created man pretty much in his present form at one time within the last 10,000 years."[1]

This disagreement does not result from simple scientific ignorance, as would be the case with a question about Einstein's theory of relativity, which is understood by a small fraction of advanced students. Nor does it derive from an ambiguity of the sort that got Pluto demoted from its prior planetary status. The age of the earth is a topic encountered in geology, astronomy, and biology, which students are already studying in middle school.

In the public high school down the street from my office, there are at least four textbooks in use in earth science alone that discuss the age of the earth. All of them present data from studies of radioactivity indicating the earth is between four and five billion years old.[2] Nowhere in the approved public-school curriculum is there *any* discussion of a ten-thousand-year-old earth. The five-billion-year estimate for the age of the earth is reinforced by science television shows, museum presentations, and plaques in national parks.

If polls reported that people did not know the age of the earth in the same way they don't know why it is colder in the winter than the summer,[3] we could simply and justifiably roll our eyes about the sad state of science education in America. But this is not what the polls indicate. Polling data suggests that half the people in America reject the scientifically determined age of the earth in favor of the age provided by the creationists. That the creationists have managed to spread their message so widely and so effectively makes you wonder if perhaps God isn't on their side, as they claim. They clearly understand how to wage a culture war.

THE "ADVENT" OF SCIENTIFIC CREATIONISM

A century ago creationism was an eddy in the backwaters of an embryonic fundamentalism. Evolutionists contributed to *The Fundamentals* and, although all the contributors affirmed that God was the creator, there was no universal rejection of evolution as a mechanism of creation. The most consistent creationist voice belonged to the new Seventh-day Adventist movement, which looked to the mid-nineteenth-century prophetic writings of Ellen White for guidance.

White was an eccentric prophetess whose writings have been more widely translated than any other American writer. As mentioned in Chapter 2, she experienced the "Great Disappointment" in 1844, when Jesus failed to appear. White nevertheless remained faithful and began receiving her own visions. Before long she was at the heart of an emerging new sect that now boasts more than fourteen million followers in two hundred countries. Her prodigious literary output exceeded five thousand articles and forty books.

Among White's influential writings is *Patriarchs and Prophets* in her series "Conflict of the Ages," first published in 1890. In this fascinating text White offers an expanded vision of Bible stories such as the Genesis creation accounts, the fall, and Noah's great flood. In a curious twist of history, modern creationism can be traced to her expansion of the Genesis flood narrative.

By the middle of the nineteenth century, when White's visions began, geologists, most of them Christian, had concluded that Noah's flood was a local affair confined to the Middle East. Its effects had been erased over time. This interpretation of the story, though not the most literal reading, was uncontroversial and accepted by most educated Christians. White rejected these geologically motivated "compromises" as inconsistent with the plain account given in the Bible. She insisted Noah's flood was worldwide and that it had produced all of the geological layers. The flood completely reshaped the surface of the earth, and the fossils testified to the cataclysmic nature of the flood. Earth history prior to the flood was completely obliterated, but the flood itself left the clearest evidence imaginable. Here is White's vision:

> The entire surface of the earth was changed at the Flood.... As the waters began to subside, the hills and mountains were surrounded by a vast turbid sea. Everywhere were strewn the dead bodies of men and beasts. The Lord would not permit these to remain to decompose and pollute the air, therefore He made of the earth a vast burial ground. A violent wind which was caused to blow for the purpose of drying up the waters, moved them with great force, in some instances even carrying away the tops of the mountains and heaping up trees, rocks, and earth above the bodies of the dead....
>
> At this time immense forests were buried. These have since been changed to coal, forming the extensive coal beds that now exist and yielding large quantities of oil.[4]

White's embellishment of the biblical narrative attracted little interest outside Adventist circles, but within the Adventist tradition her writings acquired a stature comparable to that of Scripture. Loma Linda University near San Bernardino in southern California, for example, was founded in 1905 to educate students in the context of White's visions and other Adventist distinctives.

White's interpretation of the flood became widely known outside Adventist circles through the writings of George McCready Price (1870–1963), who was born in New Brunswick, Canada, not far from my hometown. A self-taught geologist with little education beyond high school, Price was a gifted writer, amateur scientist, and tireless crusader in the cause of anti-evolution. His *The New Geology*,[5] published in 1923, was catapulted into relevance by William Jennings Bryan, who wielded its anti-evolutionary arguments in his crusade against Darwinism. A few decades later respected fundamentalist scholars John Whitcomb and Henry Morris joined forces to mainstream Price's ideas in *The Genesis Flood*. The book launched the modern creationist movement and helped convince half of America that the earth was just a few thousand years old.

Price defended a recent six-day creation, relying on the flood to provide an alternative explanation for the data that serve as the primary evidence for evolution. Evolution is supported by the observation that the fossil record shows increasing complexity over time. If Price could undermine this foundational evidence, the so-called geological column, the evolutionary theory resting on it would collapse.

The New Geology assaulted the concept of the geological column, the sequence of past epochs inferred from the stacking patterns found when layers of rock are exposed. Guides inform tourists traveling into the Grand Canyon, for example, that they can read geological history as they descend. The surface layer records the present and contains indicators such as existing plants, animals, and Coke cans along with Snickers wrappers and tabloids with stories about the travails of current celebrities. Lower rock layers provide information about increasingly older geological eras. At one level we find a fossil that is two million years old; farther down we have fossils that are twenty million years old. The pattern is clear. Traveling down is like going backwards in time, say geologists.

Price disagreed and, over the course of seven hundred pages in *The New Geology*, masterfully gathered every exception, counterexample, and questionable extrapolation used by geologists to argue that the geological column tells a believable historical tale. He says:

> This alleged historical order of the fossils is clearly a scientific blunder; for there are many unequivocal evidences to prove that this supposedly historical order must be a mistake. There is no possible way to prove that the Creta-

ceous dinosaurs were not contemporary with the late Tertiary mammals; no evidence whatever that the trilobites were not living in one part of the ocean at the very same time that the ammonites and the nummulites were living in other parts of the ocean; and no proof whatever that all these marine forms were not contemporary alike with the dinosaurs and the mammals. In short, the only scientific way to look at this matter is to say that we have in the fossils merely *an older state of our world;* and the man who wishes to arrange the various burials of these animals off in some sort of chronological order will have to invent some other scheme than any hitherto considered, for all such schemes of an alleged historical order which have been hitherto proposed are now seen to be wholly unscientific.[6]

We can appreciate the misleading character of this claim by considering how fossils are distributed and why Price disputed the conventional interpretation. The geological column he wants to dismantle doesn't actually exist anywhere. There is no place on the planet where the full geological and fossil history of the earth is neatly displayed in all its glory from primordial beginnings to the present. We would not, however, expect to find such a convenient distribution, as it would require that some local area remained undisturbed for billions of years while one layer of sediment piled atop another. Such an area would have experienced no ice age, no earthquake, no volcano, no flood, no continental drift, no meteorite, no bulldozer, and no major geological activity of any sort. Such a column would only be exposed to the steady entombment of successive generations of fossils buried in place by one unusual event after the other. The geological column is, instead, assembled piecemeal by combining local distributions. For this reason we have an undisturbed record of one epoch at the Grand Canyon, but we have to look a few miles away to see clear evidence of another era. And then we must look in some third place to find a sequence that overlaps both of them. By comparing thousands of partial records around the planet a complete history can be created.

Each partial geological record chronicles a bit of natural history, a "chapter" in the life of the earth. Lower layers typically contain fossils of animals very different from those that exist at present. Upper layers contain fossils similar to those presently existing. And middle layers contain fossils in between. By lining up these partial histories with each other a more complete record can be developed. Often the newest part of an old formation overlaps the oldest part of a newer formation, connecting them. There are many strat-egies for making these connections, one of which uses "index fossils."

Certain fossils are found often enough in the same geological layer that they can be used as an "index" to date the layer simply by their presence. By analogy, when my mother was young, Newfoundland had not jet joined Canada and was issuing its own postage stamps, some of which I have in the collection she passed on to me. A letter with a Newfoundland stamp on it belongs to that brief era after the establishment of the Canadian post office but before Newfoundland became a part of Canada. Because this history is well understood, historians can use these stamps as historical dividers, like a bookmark slid into the pages of time. In the same way, index fossils point to particular geological periods and, because such fossils have been correlated with other indicators of age, it is possible to infer from the fossil alone the age of the rock in which it appears.

Price rejects all this, highlighting exceptions called "thrust faults." Thrust faults occur when geological material gets knocked out of its normal spot. Sometimes upheavals and earthquakes invert the layers, which makes it look as though the fossils and other age indicators are in the wrong order. Other times material is pushed or "thrust" into the middle of an otherwise orga-nized stack, like the book reviews I sometimes insert into the middle of my books. Identifying thrust faults is pivotal to making sense of data that appear out of order. Price, however, suggests that the "theory of 'thrusts' is a rather pitiful example of the hypnotizing power of a false theory in the presence of the very plainest facts."[7] The reason that faults are invoked at all, he says, is "solely because the fossils are found occurring in the wrong order."[8]

Lay readers, unfamiliar with geology, often find Price's argument con-vincing. William Jennings Bryan certainly did. But informed readers are ap-palled. Why would Price make such a big deal about fossils in the wrong order? Only a tiny fraction of the rock formations have this problem. And why would Price say that "fossils ... in the wrong order" is the *only* rea-son to claim that a section of rock has been overturned? This is as ridicu-lous as arguing that "tires on top" is the *only* way to tell that a car has rolled over. When a geological formation has been inverted there are *many* indica-tors. Fossilized animals will be found on their backs, with their feet point-ing up, not likely the orientation in which they were buried. Strata with rain and wind marks will have those marks on the underside. An eroded trench might face down rather than up. An inverted formation may contain large objects with their centers of gravity high rather than low. A pyramid-shaped boulder, for example, might be found with its point down. Radioactive dating of the rock layers, which generally correlates almost perfectly with the age of the fossils, will be backward. There are many ways to identify an inverted formation.

Price's book presented many photographs of such formations. He was widely traveled and, for an amateur, well read in geology. How could he make such inexcusable errors in an ambitious textbook he hoped would overturn the entire science of geology? Who, exactly, was he writing for? He certainly was not writing for anyone with geological training; experts would, and did, immediately recognize the falsity of these claims.

In addition to challenging the central concepts of geology, Price offered his own bizarre replacement geology. Prior to the great flood of Noah, he stated with assurance, the earth was a delightful planet-wide greenhouse. He claimed that everywhere the terrestrial "climate was a mantle of springlike loveliness." Although he offered no explanation for how this climate originated, he assured readers that this floral era was, quite simply, "a matter of fact," a claim he hung on the most speculative of threads.[9] Furthermore, this global paradise was the "only" climate that existed anywhere on the earth prior to the flood. During this epoch the plants and animals were "larger and more thrifty-looking than their corresponding modern representatives." Our modern counterparts are "degenerate dwarfs." Unfortunately, we have not discovered a single human fossil from before the flood because God "buried their remains so completely."[10]

Readers may object that I have dug up a dead creationist and flogged him unfairly. Any 1923 geology book is bound to contain problems. The difference is that the successors to other geology books corrected and updated their content. Errors discovered in earlier texts disappeared from later texts, and the content steadily improved. This didn't happen with Price's eccentric "flood geology." It was simply recycled without advancing much beyond where it was when Bryan invoked it in Dayton, Tennessee.

The New Geology, for those who read it, stirred up a vigorous sandstorm across the geological landscape, until the entire science of geology looked obscure and questionable to outsiders. Price's writing style was comfortable and convincing for laypeople; his arguments were logical and easy to follow. The book synergistically combined elements that made it effective and, in so doing, provided a template for the anti-evolutionary work that followed.

These elements, used to great effect in virtually every creationist text since, included:

1 A preference for simple observations laypeople could easily understand and pass on in casual conversation or in speeches and sermons.

2 The use of glib generalizations that, though not false, ignored the nuances that a professional scientist would feel obligated to include. Such pandering to the uninformed came to characterize creationist writings, which were readily critiqued and refuted by specialists. This was to no avail, however, since such refutations occurred in publications that laypeople did not read and often involved subtleties they could not follow. As we shall see later, some peculiar creationist arguments had an enduring presence within fundamentalism and would continue to circulate long after they had been rejected, even by the creationists who first promoted them.

3 A winsome celebration of the commonsense insights of ordinary people over a "scientific establishment" blinded by its need to rationalize its preferred paradigms.

4 A consistent portrayal of the scientific establishment as inappropriately "secular," rhetorically glossed to mean "godless."

5 A pejorative stance toward "theory," implying that a theory is really just a guess based on assumptions, in contrast to observable facts. Price wrote "*a theory put to work is a hypothesis*. And hypotheses are always dangerous things."[11]

6 Nonstop ad hominem attacks on scientists, making them appear so closed-minded that nothing they claim could possibly be legitimate: "They made many and grievous blunders of observation, due to the hypnotic suggestion of their supposedly infallible theory."[12]

Despite Price's emergence as "the principal scientific authority of the Fundamentalists,"[13] he had little formal scientific training, virtually no publications in peer-reviewed journals, and no credentials of any sort beyond an introductory education to which he kept adding. He was not a member of the scientific community and, except for his notoriety as a popular enemy of evolution, was unknown in scientific circles. This was not a problem, however, for Price's audience was uninformed fundamentalists who were content to know that some smart guy had proven that the biblical flood refuted all the geological evidence for evolution, even though they couldn't remember how.

Canadian humorist Stephen Leacock captures this attitude perfectly in his *Sunshine Sketches of a Little Town,* in which he introduces Mallory Tompkins and Mr. Pupkin, who "used to have the most tremendous arguments about creation and evolution." "Tompkins," writes Leacock, "used to show that the flood was contrary to geology, and Pupkin would acknowledge that the point was an excellent one, but that he had read a book—the title of which he ought to have written down—which explained geology away altogether."[14] Pupkin's book that "explained geology away" may have been Price's *Illogical Geology,* which appeared in 1906,[15] six years before Leacock's *Sunshine Sketches.* In any event, a Pupkinesque confidence emerged within American fundamentalism that there was evidence "out there" that refuted evolution.

Price's version of White's young-earth creationism, his "flood geology," did not catch on at first. Evangelical Christians, with notable exceptions, remained content to accept the great age of the earth, inserting the geological ages into the hermeneutical orifices in the Genesis creation story. The scholarly community looked at Price with amusement, a geological Don Quixote heroically tilting at scientific windmills, convinced he could single-handedly overturn two centuries of geological work. His work was riddled with so many simple errors that any consideration he did get from the scholarly community was highly critical.

In the final analysis Price's ideas served little purpose beyond providing an "authority" for fundamentalists to invoke against evolution. Bryan and other leading anti-evolutionists certainly looked to Price as an authority. And for decades he was *the* scientific authority. Most people, however, accepted the great antiquity of the earth, content to believe that the days of Genesis were geological epochs. Leading Protestant thinkers, even within the conservative evangelical camp, were satisfied that the great flood of Noah was a local affair, and not Price's global catastrophe.[16] Evolutionary voices also were quiet in the decades after Scopes, with textbook publishers pandering to the Pupkins to avoid controversy and ensure that their sales were strong.

THE GENESIS FLOOD

On October 4, 1957, the Soviet Union launched the first Sputnik satellite. The event jolted a complacent America that had been resting on its scientific laurels ever since its atomic bombs and radar had won World War II. Immediately alarmed about the state of science and science education, the government poured money into the reform of high-school science teaching, including biology.

These reforms led to a new high-school biology curriculum, the Biological Sciences Curriculum Study, which assigned evolution a prominent role, consistent with what it was playing in the field of biology. For the first time in decades, biology texts had their evolution content driven by *scientific* rather than *commercial* considerations. Dramatically increased coverage of evolution resulted, which generated widespread and negative reactions from fundamentalists. A growing hunger for an anti-evolutionary messiah to replace William Jennings Bryan began to develop.

That messiah came in from the wilderness in 1961 carrying a book under his arm that would come to define creationism in America. In so doing it would establish itself as perhaps the most influential text on any topic in the second half of the twentieth century. This messiah and architect of contemporary creationism was a winsome Southern Baptist named Henry Morris, who got his academic start at Rice University in Houston, Texas, then known officially as Rice Institute and locally as a "hotbed of infidelity."[17] Morris attended Rice because it was free and close enough that he could live at home, his family having come on hard times in the Depression.[18] He was a brilliant engineering student and graduated Phi Beta Kappa in 1939.

Morris loved the Bible, which he studied and read on a daily basis, a practice he continued for his entire life and modeled as an important family activity for his children. He helped the Gideons distribute Bibles. Consistent with his fundamentalist commitments, he believed that God had authored the entire Bible and that all of its statements were true, whether they pertained to history, science, morals, or fishing and farming.

In the 1940s Morris returned to Rice to teach engineering. He grew increasingly committed to reading the biblical stories of the creation, fall, and flood as literal history. He wanted to validate these literal interpretations with scientific models. The important task of developing these models had been neglected because biblical scholars, even those claiming to be evangelicals, had been promoting alternative readings. In particular there were few fundamentalists insisting on a young earth, preferring the gap theory or the day-age interpretation of Genesis. Morris was especially disturbed by the near universal belief that the biblical flood had not been universal, but local, inundating Noah's stomping ground but little else. Why in the world, asked Morris quite sensibly, would Noah labor for a century to build an ark to save his family and the local animals when they could simply have migrated upland? To Morris, such a reading of the flood story was blasphemous, a compromise with a secular science that reflected an unwillingness to take God at his word.

Morris was a new and improved George McCready Price. Whereas Price was self-taught in science, with credentials easily dismissed by his critics, Morris had a stellar and relevant academic pedigree culminating in a Ph.D. in hydraulic engineering from the University of Minnesota. He minored in geology and mathematics. He could hardly have been more qualified to work on flood geology if he had been Noah's first mate. Price taught at Bible schools and other small religious colleges; Morris taught at Rice and later became the head of the engineering department at the respected Virginia Polytechnical Institute. And, perhaps most important, Morris belonged to the Southern Baptists, a mainstream evangelical tradition with none of the baggage of the marginal Seventh-day Adventist tradition to which Price belonged. If Morris was the messiah the creationist movement needed, Price had been his John the Baptist, crying in the wilderness for Christians to prepare.

The book that would launch the creationist movement and move fundamentalism strongly away from its errant and misguided reading of the Bible was *The Genesis Flood: The Biblical Record and Its Scientific Implications.* Published in 1961, the bombshell was coauthored by Morris and John C. Whitcomb, Jr., an Old Testament scholar. Whitcomb, like Morris, was well credentialed, with an honors degree in history from Princeton University and a doctorate from Grace Theological Seminary. In contrast to the gentle and diplomatic Morris, Whitcomb was angry about the deplorable state of biblical interpretation. Much of his wrath was directed at Bernard Ramm, who in 1954 had published the influential *The Christian View of Science and Scripture.*

Ramm had worked with the great Swiss theologian Karl Barth and gained a significant reputation himself. He taught at several evangelical institutions, including Baylor University, and combined a deeply held belief in the inspiration and reliability of the Bible with a respect for science. He rejected fundamentalist claims that a "high" view of Scripture demanded that all its references to the natural world be taken literally.

"Conservative Christianity," he wrote, "is caught between the embarrassments of simple fiat creationism, which is indigestible to modern science, and evolutionism, which is indigestible to much of Fundamentalism." The only way out of this "impasse" is to accept "progressive creationism."[19] Progressive creationism was the idea that God created the world and its life-forms gradually, along the trajectory disclosed in the fossil record and with methods similar to those described by evolution. Ramm took the geological record at face value, rejecting claims by Price and others that it was an artifact of Noah's flood. The earth is ancient, said Ramm, not young; and Noah's flood was local, not global. Ramm labeled fundamentalists who rejected these ideas "hyperorthodox" and accused them of various intellectual crimes from inconsistency to gross and inexcusable ignorance of science.[20] Their "pedantic hyperorthodoxy" caused the "great cleavage between science and evangelicalism" in the nineteenth century. Their continued obstinacy was only widening this gap, guaranteeing that Christianity would continue to be humiliated. Science marches onward, said Ramm, but creationism keeps running in place on its creaky nineteenth-century treadmill.

Ramm's broadsides enraged Whitcomb, who decided to take him on in print. He began revising for publication his dissertation, *The Genesis Flood,* and searching for someone approximating a geologist who would bring scientific credibility to the project. Henry Morris could not have been a better fit. Whitcomb and Morris's classic text has since been through dozens of print runs and sold hundreds of thousands of copies. What it accomplished is nothing short of astonishing and makes it easy to see why the authors claimed that God helped them write it.[21]

At the time the book appeared, most fundamentalists accepted the great age of the earth, in agreement with the scientific community. Now, a half century later, fundamentalists are largely united under the banner of young-earth creationism. In the decades that span this transformation, there were no scientific discoveries undermining the great age of the earth; no new books were added to the Bible; no advances in biblical interpretation suggested more literal readings of Genesis. And yet millions of American fundamentalists changed their minds about the age of the earth.

The Genesis Flood, in a nutshell, is two long arguments woven together, a logical double helix. The first is warmed-over Price, updated, energized, and stripped of its Adventist origins. The second is an assault on Ramm and his school of "compromise" biblical interpretation.

The Price connection is all but invisible unless one reads *The Genesis Flood* alongside *The New Geology.* By the time Whitcomb and Morris began work on their book, Price's public image was that of a geological clown, a strange one-man scientific community combing the planet for evidences to support the bizarre visions of a nineteenth-century prophetess. *The New Geology* had been blasted to bits in the press, and even those who endorsed it, like Bryan, strangely missed the fact that it was not compatible with the prevailing old-earth creationism. One can't help but wonder if its loyal promoters read it carefully. It may be that it was embraced, Pupkin-style, as an "authoritative refutation of evolution," with scant attention paid to exactly how it accomplished that feat.

In Whitcomb's early draft of *The Genesis Flood,* Morris had noted with caution that the geology was "merely a survey of George McCready Price's arguments."[22] Mindful that Price's book had flopped, Morris worried that a recycling might not fare much better. Whitcomb agreed, and they set out to recast Price's work in a way that retained its strengths but hid its origins. When *The Genesis Flood* was finally published, there were but four references to Price in the index and nothing of substance in the text itself. Morris, forever gracious, was concerned about this move and apologized to Price when he asked him to review some of the chapters that drew heavily on his work. Price was not upset, but some of his supporters felt Whitcomb and Morris were disingenuous and unprofessional in concealing their debts to Price.[23]

The Genesis Flood, however, was more than recycled Price. Its extensive discussion of biblical interpretation, for example, had no counterpart in *The New Geology.* And it addressed at least one of the difficulties in Price. Recall Price's flimsy argument that before the flood the earth was a greenhouse of "springlike loveliness."[24] Whitcomb and Morris kept this idea and went on to argue that this greenhouse hypothesis can actually be explained by the presence of a great canopy of water vapor circling the earth. This canopy, they argued, did two things: it protected the earth from harmful radiation from space, enabling preflood earthlings to live for hundreds of years and creatures to grow to gigantic sizes; and it provided the waters for Noah's flood.

The vapor-canopy hypothesis, which has been a critically important fixture in creationist thinking since *The Genesis Flood* was published, hangs on threads connected tenuously to two biblical passages. The existence of the "canopy" is supposedly implied by Genesis 1:6–7, where we read: "And God said, Let there be a firmament in the midst of the waters, and let it divide the waters from the waters. And God made the firmament and divided the waters which were under the firmament from the waters which were above the firmament: and it was so" (KJV). The canopy as a *source* of the waters for the flood comes from Genesis 7:11, where we read that "the windows of heaven were opened" to provide floodwaters.

This improvement on Price "rounded out" the flood geology model of *The Genesis Flood*. An Edenic preflood world, where people lived for centuries (as the first chapters of Genesis record) and animals grew to grand sizes, was destroyed by a global cataclysm. The disaster precipitated a vast water canopy from above and released waters from the "great deep." This flood created the fossil record we study today. The postflood environment is, of course, anything but Edenic, with unhealthy radiation from space and ongoing geological and atmospheric turmoil.

The submerged reworking of Price was accompanied by an unsubmerged assault on Ramm, who was damned on two accounts.[25] In the first place, Ramm had insulted Price, calling his ascendancy to the position of fundamentalism's "leading apologist in the domain of geology" one of the "strangest developments of the early part of the twentieth century."[26] Ramm's argument laments the abandonment of serious intellectual engagement with science in favor of Price and other pseudoscientific cranks. In the second place, Ramm proposes a less literal approach to biblical interpretation, still maintaining that the Bible was fully inspired by God and infallible. Ramm put people like Whitcomb on the defensive. The response needed to be careful and measured and not resemble the anti-intellectual "hyperorthodoxy" that Ramm claimed had come to describe evangelicalism.

History, for better or worse—actually just worse—is on the side of the fundamentalists when it comes to issues of biblical interpretation. This odd reality provided a strategy for Whitcomb and Morris to counterattack Ramm. Ramm's "questionable" views about the Bible arose from *compromises*. Positions Christians had historically held were abandoned or modified in response to secular developments. Nobody, for example, construed the days of creation as geological epochs until scientists uncovered what looked like evidence for the great age of the earth. Nobody thought Noah's flood was local until geologists supposedly found portions of the earth that had never been flooded. Nobody proposed a long developmental process for creation until scientists decided that such a process was clearly disclosed in the fossil record.

The central role of the Bible in Christianity, together with the doctrine of biblical inspiration, makes it hard to simply shrug one's shoulders and say, "I guess the Bible got it wrong there." My favorite example, which I invoke every year when I teach this material, is the Bible verse thrown at Galileo when he argued that the earth was moving around the sun and not vice versa: "The world is firmly established; it cannot be moved" (Ps. 93:1, NIV). I have yet to encounter a Christian who, when confronted with this verse, which is rarely quoted from pulpits, responds: "The guy who wrote that thought that the earth didn't move, but he was wrong. The Bible verse contains an error." (Galileo couldn't find anyone with that response either.) The usual response is to suggest that the Psalmist meant something less immediately obvious like "Human beings cannot move the earth" or "The earth is a secure place for humanity." Such responses emphasize that the text has multiple *interpretations,* and showing that one *interpretation* is wrong is not equivalent to finding a factual error in the Bible. The goal, for conservative readers, is always to look for some *plausible* interpretation of the biblical text that keeps it free from error.

There is an informal pecking order in conservative Christianity when it comes to biblical interpretation. Among people with a conviction that the Bible is without error, the "highest" view of Scripture is the one that reads the text most literally. The "lower" views are those that "compromise" the most literal meaning of the text by developing "interpretations" that, however legitimate, are not the most obvious and certainly not the most literal. Such views are considered to be "compromises." It was precisely on this point that Whitcomb crossed swords with Ramm, despite their shared convictions about the nature of the Bible.

In the introduction to *The Genesis Flood,* Whitcomb and Morris write: "We desire to ascertain exactly what the Scriptures say.... We do this from the perspective of full belief in the complete divine inspiration and perspicuity of Scripture, believing that a true exegesis thereof yields determinative Truth in all matters with which it deals."[27] In the first chapter of *The Christian View of Science and Scripture,* Ramm writes that he "believes in the divine origin of the Bible and therefore in its divine inspiration." He "emphatically rejects any partial theory of inspiration" and anticipates that, through careful scholarship, "science and Scripture will eventually concur."[28]

There is but a subtle difference between these two positions as they played out in practice. Ramm took science seriously and would use its conclusions to modify his interpretations of the Bible, while Whitcomb and Morris regarded the most natural interpretation of the Bible as necessary. They rejected any science in conflict with that interpretation. This disagreement shows up consistently throughout the Bible, but nowhere with more clarity than in the first chapter of Genesis, which describes God's creation of the world in six days, each with an evening and a morning. The most obvious interpretation of this passage is that the "days" refer to ordinary twenty-four-hour days. However, as the science of geology developed, it became clear that the developmental history described in the Genesis creation story was much longer than six days. A "pure" and uncompromising biblical literalist, with the "highest" view of Scripture, would simply reject these conclusions from geology and look to undermine the relevant science. This was the approach of Whitcomb and Morris. Ramm, however, was unwilling to set science aside and preferred instead to look for alternate interpretations. Such an interpretation could be found in the proposal that the "days" of Genesis were long periods. Thus, by choosing a less obvious but still plausible interpretation of Genesis, harmony was achieved between science and the Bible, without acknowledging error in the Bible.

The Genesis Flood is a long argument against Ramm's approach. For Morris and Whitcomb, the Bible is God's revelation to humanity. Science, in contrast, is sinful humankind's fallen and feeble attempt to understand the natural world. How can we possibly use the latter to understand the former? Should we not take God at his word and interpret natural phenomena within the framework of the biblical revelation? If we start compromising the literal meaning of the Bible to bring it into alignment with science, where do we end? Is this not a dangerous slippery slope? Will this approach not ultimately prove corrosive to faith in the Bible and, when we are finished, will we not discover to our dismay that the "acid of compromise" has eaten away the entire Bible?

Whitcomb and Morris declared war on Ramm. And they won. *The Genesis Flood,* a half century after it first appeared, is currently sales-ranked at

30,870 on Amazon.com. Ramm's *The Christian View of Science and Scripture* ranks at 1,168,000 and is technically out of print.

REFLECTIONS ON CREATIONISM

I have dwelt at length on *The Genesis Flood,* tracing its origins to two roots. The first was the flood geology of Ellen White and George McCready Price, which it embraced, albeit sheepishly. The second was the nuanced approach to Scripture of Bernard Ramm, which it rejected. These two sources form the basis for the entire 518 pages of the book.

The argument in *The Genesis Flood* is compelling to a conservative Christian layperson interested in science, precisely the sort of lad I was when I read it in high school. The lengthy footnotes on virtually every page, the constant invocation of authorities from every imaginable discipline, the diagrams and pictures, and the synergistic credentials of the authors all combine to endow the book with authority. The uncompromising respect for the complete truth and accuracy of the Bible is comforting to readers raised in homes where the Bible is read daily, memorized, and applied consistently to daily life. Christians everywhere look to the Bible, and there are no more encouraging champions of its veracity than John Whitcomb and (the late) Henry Morris. The argument itself is both compelling and intriguing: compelling because of the logical rigor and careful reasoning, and intriguing because of the suggestion that the entire scientific community had run off the rails in trying to explain all of creation without acknowledging God. There were even Bible verses alluding to this kind of "last days" intellectual apostasy.[29] Their entire presentation was very believable, which is why the argument won the day and now sits at the center of the evangelical worldview.

The Genesis Flood sold a few hundred thousand copies, which surprised everyone and probably got someone fired at Moody Press. Fundamentalist Moody Press declined the book because its editors thought that the "day-age" interpretation of Genesis was both true and so generally accepted by its target audience that it would be pointless to publish a mammoth book assaulting it. Despite relatively robust sales, however, the numbers indicate that less than 1 percent of American evangelicals actually read *The Genesis Flood.* So how did its central message fare so well?

THE PUPKINIZATION OF AMERICAN EVANGELICALISM

The Genesis Flood was a watershed event in the evangelical engagement with science. It represented the abandonment of a long tradition of taking mainstream science seriously. The founders of modern science—Galileo, Kepler, Newton—had all been deeply religious and invested in the integration of science and faith. The nineteenth-century milieu that gave birth to Darwinism had a similar set of deep Christian thinkers—Faraday, Wallace, Gray, and even the young Darwin. These scientists took seriously the task of integrating evolution with key Christian doctrines. Evangelicals in the first half of the twentieth century, from B. B. Warfield at Princeton to the infamous Bernard Ramm, had continued this task. This was all to end.

The Genesis Flood was intellectually disastrous on two fronts. On the scientific front it convinced far too many evangelicals that there was an "alternative science" out there for them, and that this alternative was consistent with a simple reading of the Bible and required no complex "reinterpretations." On the religious front, *The Genesis Flood* convinced far too many evangelicals that a faithful interpretation of the biblical text required subscribing to a young age for the earth and a worldwide flood. Furthermore, these beliefs were dramatically elevated in importance. For almost two thousand years virtually nobody made a big deal about the age of the earth or the details of the flood. *The Fundamentals,* published at the beginning of the twentieth century, paid scant attention to these topics. Now, suddenly, Christians had to embrace these beliefs in order to be faithful to the Bible they cherished. A broad range of acceptable positions on this topic collapsed into one. "Creationism" lost almost all of its traditional theological meaning and became a political label attached to Christians who reject evolution and embrace a young earth and worldwide flood. Twenty years later, when Isaac Asimov expressed his dismay by saying, "Creationists are stupid, lying people,"[30] everyone knew exactly what he was talking about. The same statement a century earlier would have been deeply ambiguous.

FIVE DECADES OF "FLOODING"

The Genesis Flood became "brand-name creationism" and created the paradigm for almost all subsequent developments in the creation–evolution controversy. When creationism appeared in the courts, it was this brand. When a "creation research" center was started, it was to explore this brand. When a "creation journal" was launched, it was to promote this brand. When a "creation museum" opened up, it was to present this brand. The shelves on Christian bookstores filled up with various presentations of this brand of creationism, from popularizations of the weighty Whitcomb and Morris tome to cartoon books for children showing people and dinosaurs on Noah's ark together. Bible study materials appeared, arguing that this is the only way to interpret the various relevant Bible passages and even that these ideas are essential for Christians.

"Research" centers and organizations emerged to support this surging creationism. The most influential was the Institute for Creation Research (ICR), affiliated with the fundamentalist Christian Heritage College in San Diego. The college had been started by Morris and Tim LaHaye, the latter of whom went on to great fame and wealth as the best-selling coauthor of the wildly popular *Left Behind* series. Morris was proud that Christian Heritage was the "first college in modern times formed in order to provide a liberal arts education based specifically on strict Biblical Creationism and full Bible controls in all courses."[31] He envisioned ICR making important contributions to "creation science" and turning into a major research center. Imagine the impact of qualified Ph.D. scientists working with eager graduate students on sophisticated creation research projects! Morris dreamed of moving the scientific community away from its exclusively old-earth, evolutionary paradigms.

Morris's dream of gaining academic respectability for creation science crashed and burned in the decades following the emergence of ICR and other "research" centers. Creation science proved unable to establish itself and made no inroads into mainstream science. If anything, by becoming better known, it marginalized itself even further and became a sad academic joke to be studied as an example of "pseudoscience,"[32] written about in books with titles like *Why People Believe Weird Things*[33] and lampooned on television shows.[34]

The key "scientific implications" outlined in *The Genesis Flood* proved incapable of inspiring meaningful research, although some efforts were made. The intriguing "vapor canopy" idea received much attention, but nobody could come up with a model to show how it might have developed or been sustained. An ICR physicist created computer models that yielded results even he described as "disappointing for advocates of a vapor canopy."[35] Claims that radiocarbon dating was unreliable led nowhere. A recent study speculatively and hopefully looked for a mechanism tied to the flood that would produce "millions of years' worth of nuclear decay ... in just days."[36]

One of the more intriguing claims in *The Genesis Flood,* accompanied by a photograph, is that human and dinosaur footprints have been discovered together in a riverbed in Texas.[37] This appeared to confirm the provocative claim that dinosaurs were contemporary with humans rather than having gone extinct seventy million years earlier. The photos were circulated broadly and appeared in countless creationist books and even in a film distributed by ICR titled *Footprints in Stone.* The claim turned out to be fraudulent. Some of the footprints had even been chiseled in the stone by a local man who then sold them. ICR eventually withdrew the film from circulation, but not before the argument had taken up residence within fundamentalism, where it is still being trotted out by charlatans and the hopelessly uninformed.

The growing popularity of creationism threatened public education and raised concerns within the scientific community. Many books and articles appeared refuting its pseudoscientific claims and defending the conventional interpretation of earth history. But such claims were lost on rank-and-file evangelicals, who were certainly not going to read books attacking their faith. They knew better than to open the pages of books by godless agnostics like Richard Dawkins and Stephen Jay Gould or even deluded fellow believers like biologists Ken Miller and Darrel Falk.[38] Critics of creationism were often rude and dismissive and appeared to have agendas that went beyond the truth of various claims about the natural history of the earth. I mentioned above that Isaac Asimov called creationists "stupid, lying people." Oxford biologist Richard Dawkins, in similar vein, stated that anyone who rejects evolution is "ignorant, stupid, or insane."[39] Tufts University philosopher Daniel Dennett suggested that creationists should be "quarantined" and their children told that their parents are engaged in the "spreading of falsehoods."[40] Such examples can be endlessly multiplied.

These famous critics failed to grasp that creationists are also committed Christians and many of them are reasonable, generous, and motivated by the noblest of intentions. Thoughtful Christians sense something disingenuous about the mean-spirited lambasting that accompanies what should be a civil argument about science. These diatribes, they reason, must derive either from a great insecurity about one's own beliefs or a sinister spirit working to undermine God's eternal truth. As mentioned earlier, Morris explored this latter thesis in a popular book titled *The Long War Against God,* in which he made the extraordinary claim that evolution came directly from Satan. The passage is worth quoting in full:

> Now if Satan (or Lucifer) is going to believe that God isn't really the Creator, then he has to have some other explanation. That's why I have to say that Satan was the first evolutionist. Evolutionists ridicule me for saying that, but again, I can think of no better explanation for how this worldwide, age-long lie came to be, than through the father of liars, who is the devil. Satan is the deceiver of the whole world, but he has deceived himself most of all!
>
> And he still thinks, apparently—because he's still fighting against God— that somehow he's going to win. So he keeps on fighting. He has to use the same lie with which he deceived himself, that the universe is the ultimate reality, that it's evolving itself into higher and higher systems, and that now men think they can even control its future evolution. Men can develop human beings and other things the way they want them in the future if Satan can just get control of everything.[41]

By these lights it is easy to understand the passions in this ongoing cultural clash. "Lying, stupid, wicked creationists" battle "satanically inspired evolutionists" to see whose version of natural history will win, whose creation story will be embraced by America. The overheated rhetoric is long past communication; it is nearly impossible to find a civil conversation on this topic anywhere. The evolutionists have won the academy, the prize being public schools, courts, and public television. The creationists have won the grass roots and created a self-sustaining (pseudo)scientific subculture with its own standards. They have their own publishing houses, magazines, colleges, and even their own accrediting agency, the Transnational Association of Christian Colleges and Schools (TRACS). TRACS requires member schools to affirm belief in the "special creation of the existing space-time universe and all its basic systems and kinds of organisms in the six literal days of the creation week."[42]

"WHAT A FOOL BELIEVES"

Henry Morris's dream of a creation-science research program gave way to a populist movement repeating anecdotes in the way that Pupkin argued with

Mallory Tompkins. The anecdotes remain sheltered in a subculture where, insulated from peer review, scholarly consideration, and scientific advance, they reproduce and thrive. When I ask my students how many of them have heard that dinosaurs were contemporary with humans, hands go up. How about carbon dating being unreliable? Hands go up. The fossil record full of holes? Hands go up. My colleagues at secular schools in conservative parts of the country report the same phenomena.

I am not surprised by this, for these are the stories of my youth, provided by preachers, Sunday school teachers, and the books I was encouraged to read. It wasn't until I started studying science in earnest that I discovered that these stories were simply rubbish. Many of these stories have even been quietly repudiated by the creationists themselves. There is little evidence, however, that the creationists care how new players come to be on their team. *The Genesis Flood,* after forty-four printings, has never been revised and still contains pictures of fraudulent footprints "proving" that dinosaurs coexisted with humans, despite the authors' appropriate disavowal of that claim.

There is no reason for anyone, Christian or otherwise, to take any of these claims seriously. The key ideas being promoted under the banner of "scientific creationism" originated in Ellen White's "visions." And the ideas might have stayed within the cloisters of the tiny Adventist sect, had not a clever amateur geologist named George McCready Price started to bang the flood-geology drum. Even Price won but few converts, and it wasn't until Whitcomb and Morris produced the masterful *The Genesis Flood* that the argument took off.

CREATION GOES GLOBAL

The popularity of scientific creationism is a fascinating phenomenon, authentically American in many ways and incomprehensible to Europeans. I was recently in Rome to address a conference at the Vatican on America's peculiar attraction to creationism. European Christians remain in dialog with mainstream science, harboring no fears that evolutionary biologists are all possessed by the devil. That half of America maintains allegiance to a set of ideas they discarded a century ago is beyond belief. Nevertheless, creationism appears to be going global and, although its influence abroad is limited, there are indications that it is time for the global scientific community to start preparing a response.[43]

Scientific creationism has climbed onto the radar screens of American intellectual culture only as a bad joke. Creationism's low point would have to be a 2006 episode of *The Simpsons,* "The Monkey Suit," caricaturing the key elements that turned creationism into such an intellectual embarrassment. When the show's popular evangelical character, Ned Flanders, wanders with his children into an evolution display at the local museum, he encounters, to his increasing horror "Man's Early Ancestors," "Indisputable Fossil Records," and "Unisex Bathrooms."

Agitated, Flanders asks the director, "How can you put up an exhibit on the origin of man and not have *one* mention of the Bible?" The director refers him to a display with a huge hand coming down from heaven and poking at the ground, out of which animals and humans pop into existence. The Doobie Brothers song "What a Fool Believes" plays in the background. Flanders seeks counsel from his pastor, who opportunistically sees a controversy that might get people back into his church. Seemingly interested only in celebrity, the pastor blackmails the local school into teaching creation.

A "two-models" video for the public school titled "An Unbiased Comparison of Evolution and Creationism" shows the Bible coming down from heaven on a beam of light, ringed by a halo. A choir sings, and the narrator intones that it was written by "Our Lord." The other book, *On the Origin of Species,* arises in flames, its title written in blood. Heavy metal music replaces the heavenly choir, and the narrator notes that it was written by a "cowardly drunk named Charles Darwin."

Lisa Simpson, the program's voice of reason, responds by teaching evolution in secret and gets arrested, recalling *Inherit the Wind.* A slick witness claiming a Ph.D. in "truthology" from "Christian Tech" testifies against "devolution," calling it "pure hogwash." The locals are impressed with the polished imposter; they boo the ACLU lawyer when they find out she is from New York.

Fifty years ago this humor would not have worked. But the success of Morris's anti-evolution crusade not only consolidated young-earth creationism as the primary option for evangelicals, but also introduced it to America's educated elite as a peculiar cultural phenomenon. That the movement is now so well known for its foibles is a sad commentary on just how completely lost Morris's original vision has become. His movement has utterly failed to provide a vital creation research program or to win back the scientific community. In fact, his flagship project, the ICR, now languishes under the uninspired leadership of his son John. The heart of young-earth creationism is now located at Ken Ham's Answers in Genesis organization, where writing cartoon books and funny songs about dinosaurs has replaced research and graduate education as the top priority for advancing the cause.

The discussion of young-earth creationism is now an in-house conversation, reaching few Americans outside of the evangelical subculture. There is no longer any chance that it will influence the courts, the public schools, or higher education. And, although this restriction will not interfere with book sales, the lecture circuit, or the popularity of creation museums, it does imply that this brand of anti-evolutionism has lost its chance to influence Western culture. It will play no role in reversing the tide of secularism that Christians have been fighting for over a century. It was precisely this recognition that motivated a charismatic law professor named Phillip Johnson to craft a different approach to fighting evolution called intelligent design.

CHAPTER 6

CREATIONISM EVOLVES INTO INTELLIGENT DESIGN

The arrival of *The Genesis Flood* in 1961 energized creationists. They started research institutes; they launched creationist journals; they published a library of books and articles. They created videos, Sunday school literature, and comic books. They built museums. When the Internet arrived, they produced Web sites. Whitcomb and Morris could hardly have envisioned the movement launched by their collaboration. And yet their project, by the most important yardstick of all—their own—has been a complete failure: creationism has had absolutely no impact on science. The flood geology they promoted so enthusiastically in their seminal manifesto never even appeared on the far horizon of mainstream science.

As of this writing, there is not a single scientist at a major university working within the flood-geology paradigm of scientific creationism. Not one. Not a single scientific paper explicitly promoting *any* aspect of this brand of creationism has been published in a scientific journal. Not even one.[1] Every "working" creationist is either at a tiny research center like the ICR or at a fundamentalist Christian school, like Liberty or Bob Jones, which do very little research anyway, even outside the sciences. The vast corpus of creationist literature, for its thousands of pages, consists almost entirely of popular-level books published by evangelical presses. The "scientific" output of creationist scholars is a modest bookcase of unnoticed semitechnical works and articles in "in-house" journals containing little more than miscellaneous sniping at poorly understood details in evolutionary theory.

To make matters worse, a substantial literature demolishing creationism has appeared. A few mainstream scientists, initially certain that something so fanciful and antiquated as flood geology would die on its own, recognized that creationism's grassroots popularity threatened science education. They began looking more closely at creationist claims with an eye toward refuting them and getting them out of the conversation. Unfortunately for the creationists, their assertions were all too easy to refute.[2] Unfortunately for the scientific community, the assertions they so carefully refuted kept appearing in book after book, like a gag candle that keeps reigniting after you blow it out. Apparently, some creationists believe there is no such thing as a wrong argument against evolution.

Geologists noted that there were portions of the planet, such as the polar regions, that clearly had never experienced a flood. If they were flooded four thousand years ago, as creationists claim, there would be some interruption in the seasonal stacking pattern clearly visible in ice cores and extending back for tens of thousands of years. Noah's ark, noted biologists, may have been adequate to carry most of the local animals in Noah's neighborhood, but it was way too small to house the vast menagerie that we now understand to inhabit the earth. So much evidence accumulated for the five-billion-year age for the earth that claims it was six thousand years old sounded no more plausible than the claim that it was flat.[3] The parade of evidence continued, until there was hardly a single creationist claim that retained even a shred of scientific credibility. Creationism's best-educated advocate is Kurt Wise, a geologist who obtained his Ph.D. under Stephen Jay Gould at Harvard University. Even Wise concedes that the scientific evidence was clearly stacked against creationism. Belief in creationism, for Wise, is *in spite of* the scientific evidence, not because of it; he stands with creationists because of his prior commitment to biblical inerrancy.[4] Two other forthright and well-credentialed creationists, Paul Nelson and John Mark Reynolds, share Wise's view and caution their young-earth colleagues to "humbly agree that their view is, at the moment, implausible on purely scientific grounds."[5] Greater intellectual condemnation would be hard to imagine. Needless to say, this is not the future for which Morris and Whitcomb hoped.

Effective critiques of creationism came from Christian scientists who, despite having personal faith in the Bible and fully endorsing the idea that God created the world, considered creationism to be absurd in the light of current science. Evolution, an increasing number of them argued, was, and is, both true and compatible with their Christian faith.[6] Often these scientists, who were surrounded by the people buying creationist literature in droves, had to face derision and even persecution from their own religious traditions. Biblical scholars and theologians were also weighing in critically, charging that the creationists were reading an anti-evolutionary agenda into the Bible, twisting its ancient wisdom to speak to a modern issue it never intended to address.[7]

There was widespread fundamentalist enthusiasm for creationism. The excitement drew opportunists to the cause, like hot-dog vendors to an outdoor concert. Preachers who knew nothing about science began pontificating from pulpits and writing books as if they were trained scientists speaking with authority on subjects like genetics and paleontology. The articulate and influential D. James Kennedy, for example, who heads the huge Coral Ridge Ministries, assured his millions of viewers that evolution is little more than an ill-begotten joke turned into an argument for atheism: "Darwin's ideas, which provoked laughter and lampoons in virtually every newspaper of his own day, and is a theory for which to this day there is virtually no reliable scientific evidence, have become the cornerstone of modern humanism."[8] James Dobson, Pat Robertson, and the late Jerry Falwell all launched attacks on evolution and assured their millions of listeners and readers that there were no reasons to take the theory seriously. Charlatans with virtually no education began to put Ph.D. after their names and claim to be "creation researchers." One of the most bizarre examples of this was "Dr." Carl Baugh, who became, and remains, surprisingly popular even after being exposed as a fraud.[9]

Baugh promotes a peculiar pet theory in which the preflood earth possessed a "firmament consisting of compressed hydrogen taking on near metallic characteristics, in the middle of a solid water formation about eleven miles above the earth." This amazing solid sphere was mysteriously immune to being shattered by incoming meteoroids. With the earth floating exactly in the middle of it, this shield bathed the planet in a "gentle pink glow," which enabled human brains to work at "maximum efficiency."[10] Baugh's claims are pseudoscientific nonsense, on a par with alien abductions, psychic surgery, and spoon bending. Astoundingly, though, it was this crackpot who appeared on a 1989 *Nova* program representing the creationist viewpoint! Baugh has a "museum" and continues to be a fixture on fundamentalist programs like Kenneth Copeland's *Believer's Voice of Victory*. He even has his own weekly show on the Trinity Broadcasting Network, *Creation in the 21st Century*, where he is referred to as the "foremost doctor on creation science."[11]

Comparable gibberish can be found in the writings of "Dr." Ken Hovind, who calls himself "Dr. Dino" and built a theme park in Florida organized around the idea that humans and dinosaurs coexisted. Hovind, an active crusader for creationism, has his own "Hovind Theory," which explains that an ice meteor caused Noah's flood. He claims to carry a $250,000 check for anyone who can show that evolution is the best explanation for origins. His critiques of evolution include such illuminating gems as: "Every farmer on planet Earth counts on evolution not happening. They count on it. It doesn't happen. People can believe whatever they want, but whenever a farmer crossbreeds a cow he expects to get a cow, not a kitten."[12]

Like Baugh, Hovind has bogus educational credentials. And like Baugh his writings are filled with pseudoscientific nonsense, supplemented with a surprising number of spelling and grammatical errors. This tireless crusader for creation has repudiated his American citizenship, refusing to pay taxes, and in 2006 was arrested and indicted in federal court on fifty-eight charges related to his tax problems. As of this writing he is in jail. Baugh and Hovind are but two of the more popular frauds in the creationist movement, which seems capable of generating and supporting an endless number of mountebanks and charlatans who make a mockery of both the religious faith they claim to serve and the science they pretend to understand.

The confident assertions of polemicists like Baugh and Hovind play well on Main Street. The rhetorical power of claims that evolutionists can't defend their own theory, even with a $250,000 incentive, makes it appear that evolution must indeed be dying. Perhaps it never had much life in the first place. It is thus not surprising that creationists love to claim that evolution is gradually being abandoned. In 1963, two years after *The Genesis Flood* began its long cultural tsunami, Morris published *The Twilight of Evolution,* the last chapter of which bore the title "The Death of Evolution."[13] A surprisingly steady stream of books making identical claims followed, even as evolutionary science became stronger and healthier in scientific circles.[14] The growth of this misperception no doubt reflected the steady disengagement from the scientific community of the creationists as they gradually stopped talking to practicing scientists and instead talked only to themselves. They convinced themselves that their position was so obviously superior that the opposition much surely be on its last legs.

By the 1990s creationism had become a scientific joke, consistently providing raw material for television comedies. When Peter Griffin, the lead character on *Family Guy,* was having his intelligence tested, he discovered that he ranked below "retarded" but above "creationist."[15]

This is not to say, of course, that creationism is dying. Far from it. The lampooning on television works only because the ideas are so widely popular. And, as polls, book sales, and the continuing popularity of "creation museums" attest, creationism shows nothing but robust political and economic health. Nevertheless, from a scientific and even broader intellectual perspective, creationism lacks credibility. And there can be no doubt that its ideas are irrelevant within the scientific community.

Furthermore, as a purely practical matter, the defeats suffered by creationism in the courts all but guaranteed that it would be making no comeback through the public schools. The Supreme Court's 1987 expulsion of creation science from America's public-school classrooms hammered the last nail into that coffin.

Had evolution finally won, at least on the legal and academic fronts? Was creationism now forever restricted to a large but purely fundamentalist comfort zone, out of the sight of mainstream science and without influence on American intellectual culture as a whole? Conservative Christian intellectuals, many of whom would admit to being embarrassed by creationism, found this disturbing. If evolution continued to own the academy, they reasoned, its pernicious naturalism would keep seeping into the intellectual foundations of all aspects of American life. After all, almost every leader graduates from a college or university dominated by evolutionary thinking. As these leaders take their places in positions of power, so the evolutionary thinking they imbibed in the academy would inform and control decision making everywhere. The acidic philosophy of evolution was corroding everything in America, from the Supreme Court, to foreign policy, to corporate finance, to the curricula in the public schools, to Hollywood scripts.

REASON IN THE BALANCE

No one was more disturbed about all this than the colorful, opinionated, and theologically conservative law professor Phillip Johnson, who was about to explode onto the scene like a promised messiah to rally the demoralized faithful. Johnson, a tenured professor at Boalt Hall, the law school of the University of California at Berkeley, was convinced that the issue was not evolution per se, but rather the pervasive and dogmatic *naturalism* of science. Johnson objected to the way the scientific community insisted that *all* explanations for *all* phenomena at *all* times present and *all* times past must be purely naturalistic. This effectively ruled out the *possibility* that God might be a relevant part of a comprehensive understanding of the world. Even substantial and compelling evidence would be inadequate to infer that God was involved in the world. Scientific explanations were allowed to invoke only natural laws and ordinary events for their explanation. If the face of Jesus appeared on Mount Rushmore with God's name signed underneath, geologists would still have to explain this curious phenomenon as an improbable byproduct of erosion and tectonics. A choir of heavenly angels singing carols in the sky over the White House would likewise have to be explained as an anomalous weather pattern, a flock of unusual birds, or a publicity stunt by Pat Robertson gearing up to run for president. Invoking God to explain *anything* was simply not allowed in science, *no matter how compelling the reasons for doing so.*

But why, asked Johnson, should an explanation invoking God be ruled out before even being considered? Was this not "stacking the deck" in favor of atheistic naturalism? Are the explanations provided by science really the "best" explanations? Or are they simply the "best that can be had without invoking God"? What kind of twisted logic was this?

Applied to origins, this restrictive naturalism excluded God from any involvement whatsoever in how things came to be the way they are. From the big bang, to the appearance of our solar system, to the origin of life, to the evolution of our complex brains, to the emergence of our sense of morality, God was simply not there. Or, if God was there, he was just watching, cheering from the sidelines like a fan at a football game who, although interested in the game, is irrelevant to the outcome. Everything happened by itself. Johnson found these conclusions unacceptable and began developing a strategy to level this highly sloped playing field. He would rehabilitate the argument from design and give the courts something they could not summarily reject as a breach of the battered but still standing wall between church and state. In so doing, he would single-handedly reenergize anti-evolutionism and breathe life into a new species of creationism, which he labeled *intelligent design.*

THE WEDGE OF NATURALISM

Johnson's career as the leader of this emerging "intelligent design" (ID) movement was both launched and secured with the publication of *Darwin on Trial* in 1991. The book was short, popular, readable, and rhetorically powerful. It assaulted evolution but presented no religious or biblical alternative. And, just as Price and Whitcomb and Morris had launched manifestos that guided and defined anti-evolution earlier, *Darwin on Trial* became the manifesto for the fledgling ID movement.

Johnson's stature as a respected legal scholar and the cleverness of his attack on evolution gained him entry where previous creationists had failed. No less a luminary than the late Stephen Jay Gould reviewed *Darwin on Trial* in the prestigious pages of *Scientific American*.[16] Despite the negativity of Gould's review, its mere appearance signaled the engagement of the scientific community. Gould's review also shone a national spotlight on Johnson, who was only too willing to put evolution and scientific naturalism on trial. Like the Pied Piper marching through the streets of Hamelin with children in tow, Johnson soon found himself at the head of a tiny but determined army of highly disgruntled, newly resurrected, and occasionally brilliant anti-evolutionists.

As befit a lawyer preparing a case, Johnson had a set of strategies to fight evolution. He knew there were many constituencies separately opposing evolution but on different pages regarding the biblical and theological aspects of creationism. Their differences on "minor" questions like the age of the earth or the extent of the flood had them squandering their energies on in-house quibbles or preaching to tiny choirs. They were all fighting their own individual wars against evolution, and in that odd phenomena that Freud called the "narcissism of small differences" they were energetically fighting each other. Under Johnson's leadership, they set aside their differences and joined forces against their common enemy, united in their conviction that evolution was false, while agreeing to disagree on how creationism should be understood.

Johnson's strategy contrasted strongly with that of Whitcomb and Morris, who had intertwined science and religion to great effect. But it was precisely this intertwining that ultimately resulted in the barring of their brand of creationism from the public schools. The young earth, the worldwide flood, the rejection of evolution between "kinds," the separate ancestry for humans and apes—all these ideas connected so tightly to the Genesis story that there was simply no way to "secularize" this creationism for presentation in the public schools. The creationism rejected by the Supreme Court was so obviously based on the Genesis creation story that even the articulate

Wendell Bird could not get the justices to see the science without also seeing the religion.

Johnson highlighted the naturalism of science as a philosophical problem, a bogus antireligious assumption masquerading as a scientific inference. *Darwin on Trial* concludes with sweeping, if undocumented, generalizations offered with breathtaking confidence. The "purpose" of evolution, Johnson writes on the last page, is not to understand the development of life on this planet, but "to persuade the public to believe that there is no purposeful intelligence that transcends the natural world." This assumption creates an intellectual straitjacket that prevents scientists from even *considering* possibilities inconsistent with "strict philosophical naturalism."

"Darwinists," Johnson writes, "took the wrong view of science because they were infected with the craving to be right." They confuse their "pseudoscientific practices" with real science, because they are too dense to even recognize that they have become slaves to the "philosophical program of scientific naturalism." If only Darwinists could see this with the same clarity as Johnson, they could toss off the "dead weight of prejudice" and be free at last to "look for the truth."[17]

Johnson's anti-evolution polemic made him the poster child for ID. Creationists of all stripes joined hands under the big tent he was erecting. This strategy was in evidence at a conference held at Biola College in Los Angeles in November 1996. A beefy anthology titled *Mere Creation: Science, Faith, and Intelligent Design* resulted.[18] Some creationists, of course, are uneasy about the way that ID asserts its independence from the Bible, but are nevertheless happy to join the ID crusaders in their war against evolution. The enemy of my enemy is my friend.

The foreword to *Mere Creation* celebrated contributions from Roman Catholic, Eastern Orthodox, and Jewish scholars, not to mention a spectrum of conservative Protestant thinkers—and even a disciple of Reverend Sun Myung Moon. Some contributors believed the earth was ten thousand years old; others accepted the conventional scientific age of four and a half billion. Some participants rejected evolution completely; others accepted it as long as God was constantly and intimately involved. About a third were practicing scientists or engineers. The others were philosophers, theologians, journalists, and writers. Two convictions united them: conventional evolutionary theory was wrong; and living organisms displayed clear evidence of intelligent design that could not possibly have been produced by natural processes. This was pure, or "mere," creationism, uncontaminated by divisive religious commitments or nuances of biblical interpretation. Johnson's strategy was working.

The rapid emergence of ID as a cultural phenomenon has been nothing less than astonishing. There have been many strong reactions, from sympathetic,[19] neutral,[20] and hostile[21] perspectives. ID was on the front pages of America's leading newspapers and being praised in the White House.

ID, like the creationism it was intentionally replacing, rode the same wave of anti-evolutionism that took creationism to the Supreme Court. This time, however, the energy came from a more intellectually sophisticated demographic. Flakey, fringe creationists like Carl Baugh and Ken Hovind were nowhere in sight.

The Seattle-based Discovery Institute provided generous financial support, embracing ID as a partner in its mission of cultural renewal. Much of the money came from Howard Fieldstead Ahmanson and his wife, whose affiliation with a variety of extreme right-wing causes has made some people nervous about the ultimate agenda of the Discovery Institute.[22] Proponents of ID, some of them fully funded by the Discovery Institute and devoted full-time to the cause, found a ready audience for their books and widespread demand for public appearances. Venues like public television, National Public Radio, the *Wall Street Journal,* and *New York Times* took notice, sometimes favorably. Leading science magazines gave them an occasional, if generally negative, nod. The roster of ID enthusiasts, or at least public figures coming out in support, included important national figures, from politicians like George W. Bush, Bill Frist, and Rick Santorum, to media personalities like Ann Coulter, Pat Buchanan, Chuck Colson, and Bill O'Reilly, to televangelists with huge audiences like the late D. James Kennedy.

ID was, of course, similar in many ways to creationism. As we saw in a previous chapter, the judge in the Dover case ruled that it *was* creationism, deceptively packaged to look like something else. ID's leading lights were all conservative Christians who wrote primarily for Christian audiences arguing that ID was critical for restoring God to the center of the Western worldview. William Dembski, perhaps ID's leading theorist, even titled one of his books *Intelligent Design: The Bridge Between Science and Theology.*[23] And certainly the ID movement's eagerness to join forces with the more credible young-earth creationists like Kurt Wise and Paul Nelson guaranteed that there would be a strong creationist tinge to everything under the ID umbrella.

The central argument of ID, however, was simply that the world had more design in it than could be accounted for by purely natural explanations. Dembski, ID's most prolific author, put it like this: "There exist natural systems that cannot be adequately explained in terms of undirected natural causes and that exhibit features which in any other circumstance we would attribute to intelligence."[24]

THE SEDUCTIVE POWER OF DESIGN ARGUMENTS

So what exactly are the design arguments? And what makes them so compelling? Design arguments are, in fact, logically attractive on many levels, practical, scientific, and even religious. They "feel" right, as if they somehow *have* to be true, and therein lies their attraction. And the ID proponents are indeed correct that people make judgments all the time about design. Why should such inferences be excluded from science?

An all too familiar example is the September 11 attack on the twin towers of the World Trade Center. When American Airlines flight 11 hit the north tower at 8:45, many people concluded that a terrible tragedy had occurred, but that the event was random. Eighteen minutes later United Airlines flight 175 hit the south tower, and everyone immediately *knew* a nefarious plan was being executed: an "intelligent design." Drawing design conclusions like this, as they say, is a "no-brainer," and one hardly needs specialized training to do this. Every day we routinely make such design inferences. A chocolate bar on the sidewalk is a random event; the chocolate bar I put in my daughter's lunchbox is an intentional act. The shape of a river is a random meander; the shape of an *s* is a design, even when they look similar.

Design inferences are also drawn in science. Archaeologists must decide if something they dig up is a human artifact. An interesting arrangement of stones may be simply interesting, or it may be the work of intelligent creatures. Competence in the social sciences requires the ability to distinguish between intelligent causes and purely natural ones. Astronomers listening for extraterrestrial signals believe an intelligent signal will be distinguishable from the background noise in which it is embedded.

Theologically, intelligent design in some form is almost a requirement for Christians. If God created the world, then the creation is the consequence of an intelligent act. Religious language is filled with allusions to the intelligence of God. How often do we hear that "God has a plan for your life" or "God's will be done, on earth as it is in heaven." Countless biblical passages allude to God as the source of order and rational structure. The gospel of John opens with a sort of hymn praising Jesus for being the "logos" of creation. *Logos* is a Greek term with no exact analog in English. It is usually translated "Word," but it embodies the idea of "rationality," "order," or "logic." The imagery in the Genesis creation account is of God hovering over a formless void and bringing order to the chaos. To believe in God is to believe in design.

There are three distinct strands of design: practical, scientific, and religious. These arguments have long been braided together by religious believers into a compelling argument for the existence of God. We saw earlier how William Paley created a powerful apologetic argument based on design in nature. This argument, which haunted Darwin as he gathered his observations, would eventually give way to the theory of evolution.

ID wants to rehabilitate Paley, or at least undermine the arguments that did him in. The claim that complex and interesting natural phenomena reveal the handiwork of God is indeed compelling and perennially attractive. However, although I wish it were true, it must be rejected.

WHY THE ID ARGUMENT FAILS

How can I reject the ID argument, while, paradoxically, wishing it were true? Let me start with the reasons why I, and all Christians for that matter, should wish it were true.

Like so many people, I believe in God and have done so for my entire life. And, like most believers who go on to earn advanced degrees, I have been forced to recognize that belief in God is not a simple matter. Many of the arguments that worked so well for me in high school have since lost their power to persuade. And I have a great appreciation for the counterarguments for God's existence. I understand how honest thinkers and seekers after truth like Daniel Dennett and Michael Ruse can end up rejecting God. Like that of most thinking Christians, my belief in God is tinged with doubts and, in my more reflective moments, I sometimes wonder if I am perhaps simply continuing along the trajectory of a childhood faith that should be abandoned.

As a purely *practical* matter, I have compelling reasons to believe in God. My parents are deeply committed Christians and would be devastated, were I to reject my faith. My wife and children believe in God, and we attend church together regularly. Most of my friends are believers. I have a job I love at a Christian college that would be forced to dismiss me if I were to reject the faith that underpins the mission of the college. Abandoning belief in God would be disruptive, sending my life completely off the rails. I can sympathize with Darwin as he struggled against the unwanted challenges to his faith.

If I could convince myself that ID were true, I would have a solid *argument* to believe in God, to keep those nagging doubts at bay. I would be less controversial at the college where I teach, able to affirm my students in their confident but unexamined beliefs that evolution is untrue. The president of the college would not have to worry about fundamentalist donors who won't support their alma mater because I teach there. I have many solid reasons to embrace ID and have been at times, in the words of that ancient hymn, "almost persuaded." So, when I say that I reject ID, I say it with pangs of regret. I truly wish it were true.

From my perspective, ID must be rejected on two completely separate grounds. In the first place, ID doesn't work scientifically. As I pointed out in the earlier chapter on Darwin, ID was once a viable paradigm in science, accepted by everyone. But it was not abandoned because scientists wanted to get God out of the way, as is so disingenuously claimed by some. ID was discredited because it proved inadequate as an explanation for so many phenomena. In the second place, ID is theologically problematic. To suppose that there are various structures in nature specifically designed by a transcendent intelligence, which we all know is God, is to open a Pandora's box of problems, not the least of which is the problem of bad design. And even when the design is good, what do we make of ingenious designs employed for sinister purposes?

The ID argument is simplicity itself, which accounts for its enduring and widespread popularity. We find something interesting in nature—the eye, the blood-clotting mechanism, the opposable thumb, the immune system, the bat's radar, the brain—that exhibits more complexity than science can explain. The human blood-clotting mechanism, for example, needs more than twenty different proteins to work properly; if some of them are missing, the process doesn't work. How did nature develop this complex process? It is inconceivable that all the parts just randomly came together. But it is also inconceivable that blind evolution, with no sense of where it was "going," slowly tinkered with molecules and ended up with this complex and

remarkable process. So are we not then forced to invoke an outside intervention to account for the complex design of the blood-clotting mechanism?

Leading ID theorist William Dembski developed what he calls an "explanatory filter" to determine when "design" should be invoked by ruling
out the less interesting alternatives. Applied to the example above, it works
like this. We start by asking if the blood-clotting mechanism is something
that necessarily *had* to happen, like rocks sinking in a river. The answer, of
course, is no, which means that we cannot explain the blood-clotting mechanism as the *necessary* result of a law of nature.

We then ask if the mechanism is too complex to have resulted from
chance. If the pattern is simple, like a row of three stones, it can easily be
the result of chance. The blood-clotting mechanism, of course, is too complex to be produced by chance. But not all complex patterns require intelligence to explain them. Certain complex patterns, like a cloud resembling
Homer Simpson, can indeed be produced by chance, as anyone who has
ever watched clouds on a lazy summer day can attest. But even though such
patterns are complex and interesting, they are not specified in advance.
Given that there are thousands of patterns that would look interesting, it is
unremarkable that we find one on occasion.

The final step in the explanatory filter is the question of *specification*. Is
the complexity of the mechanism under consideration something that would
have to be specified in advance? Certainly a series of typed letters explaining
how to assemble a bookcase is highly specified and the same number of letters in a random order is not. The blood-clotting mechanism is not simply
complex, but complex in such a highly specific way that, if the explanatory
filter is to be trusted, forces us to conclude that the mechanism is the result
of design.

Dembski's design-detecting explanatory filter has received much criticism, and Dembski has responded to some of his critics. Readers interested
in this discussion will find more than enough of it on the Internet.

My reaction to Dembski's filter, and to the general comments made by
other ID people about how science might detect design, is that they seem
strangely incompatible with the way that science actually works in practice. When lawyers, mathematicians, philosophers, and theologians start
pontificating about how empirical science is supposed to work, they do so
as *spectators*, not *practitioners*. Such spectators are easily confused about
the so-called rules of science. There really aren't *rules* in science. Rather,

there exists an ongoing tradition in which certain productive *approaches* become standard practice because they have proven to be helpful in generating new knowledge. Scientific approaches that are not effective in generating new knowledge are abandoned, but not because they cannot withstand philosophical or logical scrutiny, which scientists don't care about. Scientific theories without effective explanatory power and unproductive scientific approaches are rejected because they are *useless*.

The ID theorists are hung up on the so-called rules of science. But, as anyone who has earned a Ph.D. in a scientific discipline can tell you, there is no course of instruction in the rules of science required for its practitioners. You learn by becoming a part of a tradition, building intuitions, and studying under mentors. In this way, science is different from, say, law, where rules dominate and the prevailing philosophy is that following the rules leads to the truth. That is why the offices of lawyers are lined with expansive bookcases, and the law degree hanging on the wall is the product of just three years of study. Learning rules and reading books are easy. By contrast, a Ph.D. takes, on average, about ten years of specialized study. Mastering a subtle tradition of learning is complex and not something to be understood by simply looking in the window of a research lab or spending a quiet evening curled up with *An Idiot's Guide to the Scientific Method.*

If there were more historians of science in the ID movement, I think this would be better understood. The history of science is, in many ways, the history of the gradual and reluctant abandonment of ID as a helpful approach to understanding the world. Let me offer one of the clearest examples.

Isaac Newton's theory of universal gravity explained many things about the motion of the planets, including the elliptical shape of their orbits and how the size of their orbits related to their speed as they went around the sun. But his remarkable theory offered no insight into why the planets all went around the sun in the same direction. Newton was impressed by this "design." There was no law specifying this order. And it would be a strange and improbable coincidence for this to be the case. So Newton, the greatest scientist of his age, concluded that, as there was no natural explanation for the order in the solar system, it was the work of God. He "filtered out" naturalistic explanations and inferred "design."

At the time Newton made this design inference there was no satisfactory explanation of the origin of the solar system. A century later such a theory was developed and, lo and behold, the theory stated clearly that all the planets should revolve about the sun in the same direction. If Newton had known or discovered this, he would never have attributed the uniformity of the planetary directions to God.

At the time of Newton, science—then called *natural philosophy*—was a small and simple enterprise, still finding its way and fighting to be born. The world was mysterious and, with a few small exceptions, so far beyond the grasp of science that God was regularly invoked in the face of many mysteries. Newton's invocation of God is of interest precisely because of the simple clarity of the reasoning: a solid theory explained many things; the unexplained residue—the explanatory "gaps"—were attributed to God. But then science advanced and a natural explanation was discovered for the phenomena of interest, closing that gap in our knowledge.

As science advances, these gaps close. In fact, the closing of such gaps is what we mean by the advance of science. Gaps are the shadows where ignorance hides from the light of science. Inserting God into these gaps has proven, historically, to be a fool's errand and ultimately both unnecessary and embarrassing. Again and again science has made surprising advances that have allowed us to revisit these gaps in our knowledge and, often to our great surprise, close them. Historians of science know this only too well, which may be why this critically important group is so underrepresented in the ID Movement.

The central role of scientific ignorance is hidden in Dembski's explanatory filter. When theorists decide that some phenomenon is "contingent" (which means there is no law requiring it to be the way it is), what they are really saying is that they don't *know* of any such law. Newton did not *know* that solar systems form with all the planets going in the same direction, so he assumed this phenomenon was contingent and ended up invoking a design explanation.

The proponents of ID are, of course, aware of this basic objection, and they have some elaborate responses that are simply not relevant to the actual practice of real science. I don't think they have a good feel for how the historical practice of science has gradually generated a "conventional wisdom" or "common sense" that leads practicing scientists away from such explanations. When a certain approach has failed so many times, it is not irrational or dogmatic to suspect that it will fail again—it is just prudence.

The naturalism of science is like the naturalism of plumbing. When plumbers seek to understand plumbing phenomena, their experience leads them to pursue certain promising possibilities. They look for leaks, broken valves, and blockages, because this approach has been effective in the past. Why are the ID theorists not calling down the wrath of God on the plumbing community for its blinkered adherence to pure naturalism?

In the final analysis, from a scientific point of view, there is no difference between Newton's unknown mechanism for planetary directions and the currently unknown mechanism for the origin of the blood-clotting mechanism. And, I might add, the plumber's unknown leak. Why would we invoke a supernatural explanation for any of these?

The publicity surrounding the creation–evolution controversy can easily blind us to the reality that the majority of work in science has absolutely nothing to do with origins and thus couldn't make use of ID, even if it wanted to. The naturalism that ID is so quick to condemn actually works quite well, without controversy, and with the blessing of the ID community in most areas of science. Chemists make new molecules; geologists develop models to predict earthquakes; climatologists work on weather. ID enthusiasts with day jobs in science labs work comfortably within a fully naturalistic framework. Even if the ID approach were fully embraced by the entire scientific community and enthusiastically applied wherever possible, almost nothing would change outside of those small areas of biology and cosmology that study origins.

The second reason I reject ID is theological. I think ID makes dangerous and incoherent claims about God that create far more problems than they solve.

We must start by looking at ID's peculiar claim that the identity of the "designer" is of no consequence. This contrasts with the notions of Paley, for whom ID was an argument for the God of Christianity. ID theorists have actually suggested that the inferred designer could be "space aliens from Alpha Centauri; time travelers; or some utterly unknown intelligent being."[25] Although this might be true in the narrow sense that there is no way to refute the claim that a space alien designed the blood-clotting mechanism, nobody is making this claim. In contrast, virtually all of the ID people are on record enthusiastically proclaiming that God is the designer. In the subtitle of one of his books, Dembski calls ID a "bridge" between science and theology, but nowhere has he, or any other ID enthusiast, suggested that ID might be a bridge between science and space aliens.

So, what happens when we open the doors of science to supernatural design explanations? What results when we invoke God as the cause of complex phenomena like the blood-clotting mechanism, our brains, or bat radar? A lot, unfortunately.

In the first place, the God of Christianity has to be way more than just a *designer*. Centuries of Christian reflection on the nature of God have highlighted various characteristics of God: justice, love, goodness, holiness, grace, sovereignty, and so forth. Are not compassion and grace far more central to understanding God than design? Although nobody except TV preachers speaks with too much confidence about the nature of God, there is general agreement among theologians that God must be understood as multifaceted. And, although "designer" can certainly be one of these facets, it takes a backseat to God's other attributes such as love, wisdom, and grace.

Spotlighting design in nature and attributing it to God raises troubling questions. We saw how Darwin wrestled with this as he wondered why God would have endowed creatures with ingenious capacities to inflict pain. Nature, as Tennyson wrote, is "red in tooth and claw." And many of those teeth and claws are extremely well designed. Some of them would make it with flying colors all the way through Dembski's explanatory filter. However, if we run all of nature's marvelous devices through this filter, some uncomfortable results appear. The remarkable blood-clotting mechanism has received much attention and is an example that works well for the ID Movement. We are fortunate that our blood clots and have no difficulty believing that God intelligently designed the process. And we can't use our blood-clotting mechanism to inflict any pain on other creatures. So it seems like a win-win; we get something we need and God gets the credit for designing something truly beautiful as well as complex.

But what about the Ichneumonidae, which troubled Darwin? Its remarkable design would certainly make it through Dembski's explanatory filter, and we would have to conclude that the instincts of the Ichneumonidae meet the criteria for ID. So here we have an insect laying eggs inside a caterpillar. The newly hatched parasites live inside the caterpillar, consuming its internal organs. And, in a most amazing illustration of intelligent design, the Ichneumonidae eat the internal organs in a specified order that keeps their host caterpillar alive as long as possible. The parasites are born knowing how to do this; they come into the world with a genetically programmed instinct to consume the internal organs of their host caterpillar in a specific order. Forget blood clotting! This is real design. But it's horrible, as Hollywood directors know only too well. I suspect that if the champions of ID were to highlight the most revolting examples of design in nature, the evangelical community would lose all interest. Who wants a bridge from theology to a spacecraft filled with hungry parasites?

Nature is a complex web of interconnected systems. Organisms feed on each other. Parasites live within hosts. There is cruelty and barbaric behavior. There are ingenious devices for stabbing, poisoning, paralyzing, decapitating, and biting. Many of these devices appear to be designed. Are we going to run them all through Dembski's filter and ascribe them to God if they pass? How can this possibly be theologically helpful? This bridge, as Darwin figured out a hundred and fifty years ago, is better left unbuilt.

Much of nature exhibits impressive levels of design. But so do torture chambers, gun factories, and liposuction machines. Design, even intelligent design, is not automatically desirable. Promoting "design" in isolation from God's other attributes is a dangerous and ultimately self-defeating way to get God back into science. Christianity will be far better off if ID fails.

NORENE'S KNEES

Even if we can somehow convince ourselves that "intelligently designed mechanisms for doing terrible things" should be explained as the handiwork of God, there is an even more serious problem: bad design. As soon as we begin to "review" nature's many intricate mechanisms, we discover, like any reviewer, that our subject matter varies greatly in quality. Some things are designed very well and are so ingenious that we cannot help but marvel at them. Others are designed so poorly that we can only shake our heads.

Take Norene's knees, for example. Norene is a friend of mine, not yet at retirement age, who recently had both of her knees replaced. They weren't damaged by any accident or overuse in some jarring athletic activity; she wasn't even an occasional jogger. Her knees, like those of so many other humans like her, just wore out through normal usage, and long before she was through using them. Why does this happen?

Our knees are strangely designed and destined for injury. Below them are ankles that can move in several directions; above them are hips attached via "ball joints" that also permit a wide range of motion, as anyone who has ever played with a hula-hoop knows. But the knees themselves bend in only one direction. No engineer would put three joints in a row and constrain the middle one in the way our knee is constrained. The design is so bad that countless athletes have to wear a special brace to help prevent a knee from bending about the wrong axis.

The human body is riddled with design problems. Our spines are mechanically configured for walking on all fours. But we walk upright. The result? Back problems. Our mouth is designed to admit both food and air, which allows food to go "down the wrong pipe," leading to choking. We would fire an engineer who designed a car with a single opening for both oil and gas and a complex valve to keep them from mixing. There is a reason why cars are not designed like that. Why are we?

A standard roster of similar design problems is highlighted in just about any human anatomy text. We have appendices that need to be surgically removed; our mouths are too small for our wisdom teeth; our eye has a blind spot. Women's pelvises are not designed to give birth to babies with standard-size heads. And these are just a few of the specifically *human* problems. There are upland birds with webbed feet that never go in the water. Hens have genes to produce teeth that are never turned on, unless artificially induced. Genes contain meaningless sequences of "junk DNA."

There is a substantial literature looking at these aspects of nature, starting with Darwin's own reflections in *On the Origin of Species*. As the inheritor of a tradition shaped by Paley's natural theology, Darwin expected to find clear evidence of God's providential design everywhere he looked in nature. Instead, he found a globe full of exceptions, many of them quite disturbing.

There is great design in nature, to be sure. And much of it is extraordinary. But there is simply too much *bad* design to infer safely that nature's many contrivances are the handiwork of God. Our blood may indeed clot in remarkable ways, but our poor knees don't bend the way they should. Just ask Norene.

BEWARE THE GAPS

These objections I have raised to ID are far from unique. I would like to come up with a completely original critique and get everyone discussing my new insight, but that isn't going to happen for one obvious reason: this controversy is almost two hundred years old. ID was a "live" question in 1831 when Darwin boarded the *Beagle.* Naturalists at that time were trained to see "intelligent design" everywhere they looked. The Darwin of the *Beagle,* like most religious believers, myself included, had every reason to want ID to be true. After all, it provided solid scientific reasons for believing in God, carrying some of the burden of faith.

The doubts that Darwin developed about God's relationship to the natural world troubled him for his entire career. He wanted the traditional view of God as creator to remain intact because of the security that belief provided. He did not want science to be secularized. In the same way, Phillip Johnson, William Dembski, Michael Behe, and their colleagues in the ID movement desperately want God to retain the traditional role as creator, involved in enough of the details to leave divine fingerprints on nature. Like Darwin and the other Victorian mourners at God's funeral, they don't want science secularized. And so they keep fighting the young Darwin's battle for him, picking holes in this or that argument, struggling heroically and with great ingenuity to find examples in nature for which supernatural explanations can still be invoked.

When fish are removed from water and placed on land, they flail about vigorously, unable to get the oxygen they need. Fish belong in the water. Intelligent design is a nineteenth-century argument, flailing about in a new century where it doesn't belong. What looks like vigor is simply the last gasp of a way of understanding the world that died a hundred and fifty years ago.

The world is a complex place, and there is much about the universe that we still don't understand. We are centuries away from closing the many gaps in our current scientific understanding of the natural world. For a time, perhaps a long time, we may take some comfort in supposing that God hides in those gaps. We can develop ingenious explanatory filters to buttress our confidence that God is in those gaps. But it is the business of science to close gaps, and it has long been the central intuition of theology to find a better place to look for God.

Evolution, however, speaks to that all-important question of what it means to be human. And, although it may indeed be a robust science in the narrow sense of that word, when it speaks about what matters most, it does so with a deeply ambiguous voice. Different people hear different things, and in those differences reside profoundly incompatible worldviews.

CHAPTER 7

HOW TO BE STUPID, WICKED, AND INSANE

Evolution is the most culturally complex and controversial idea in all of science. Nothing else comes close. More than a century after Darwin's *On the Origin of Species,* the theory arouses hostile reactions in everyone from clueless high-school students to TV preachers to the well-educated senior fellows at the Discovery Institute. Less than half the country agrees with the scientific community that evolution is the best explanation for origins.

Courts have had to protect the central role played by evolution in high-school biology. If popular consensus refereed the schools, the embattled theory would be long gone. Teachers in school districts from Oregon to Florida struggle with how to present evolution to their students. Many don't bother, omitting or glossing over the topic to avoid controversy.[1] Some Christian colleges and universities, even accredited ones such as Cedarville College in Cedarville, Ohio, and Liberty University in Lynchburg, Virginia, teach that evolution is false.[2]

Professors at secular universities in conservative parts of the country report that students arrive in their classes with strong creationist sympathies, and many of them graduate without changing their minds. Consider the remarkable case of Kurt Wise, the leading young-earth creationist we met earlier. Wise completed an undergraduate degree in geophysics at the University of Chicago and then went on earn a Ph.D. from Harvard, working for the late Stephen Jay Gould. Wise graduated from Harvard with the same young-earth creationist beliefs he had entered college with. Creationism can be hard to dislodge.

Teaching evolution is almost impossible. In no other subject, even outside of science, is the primary challenge whether the students *believe* what is taught, rather than *understand* what is taught. Despite the simplicity of

Darwin's equation-free theory with its winsome stories of giraffes stretching their necks to reach the top of the fruit trees and peacocks preening to impress the peahens, few high-school students seem able to learn it. Despite its universal presence in high-school and college classrooms, Americans reject evolution with the same enthusiasm today as in previous decades. And despite its increasing relevance to research in biology, well-educated anti-evolutionists continue to oppose it.

The controversy surrounding evolution generates enormous press. Books appear daily attacking the theory or defending it against attack. A secondary literature has emerged analyzing the controversy and tracing its roots. Books arguing that evolution is incompatible with Christianity[3] counter those arguing the opposite.[4] There are magazines devoted to promoting evolution,[5] disputing it,[6] and even dealing with the disputations.[7] Publications nominally covering the intersection of science and religion provide disproportionate coverage of the creation–evolution controversy.[8] Television presents the same coverage. The seven-part PBS series *Evolution* devoted an entire episode titled "What About God?" to the controversy.

The creation–evolution controversy is only, in the most trivial sense, a scientific dispute. It is, instead, a culture war, fought with culture-war weapons by culture warriors. Facts are almost irrelevant. Truth is valued when it serves a purpose and not for its own sake. Name-calling, caricature, cover-up, and hyperbole dominate. Compromise is out of the question. And, in the midst of all this, high-school teachers are supposed to teach evolution to their students, oblivious to the gunfire outside the window.

A TALE OF TWO WORDS

Decades of reflecting on the evolution controversy convinces me that the conflict is only tangentially scientific. Those who would adjudicate this dispute by appealing to science are wasting their time. The conflict is not about determining the proper inferences to draw from fossils, genes, and comparative anatomy. The conflict resides at the much deeper and far more important level of *worldview*. It centers on one simple question: Can there be any role at all for God in our own creation story?

This is a far more important question than whether Darwin's theory is true. The attachment of this question to the creation–evolution controversy raises the stakes. If accepting evolution means abandoning belief in God as creator, then evolution should be opposed. And opposed with the same fervor that animated the great martyrs of the church as they marched serenely to their deaths, confident they were doing the will of God. On the other hand, science has made great strides in explaining the natural world without invoking the supernatural, and those gains must be protected. If conceding a role for God in creation turns back the clock of scientific progress, then that must be opposed in honor of Galileo, Newton, and those who fought so valiantly to create intellectual space for natural explanations of natural phenomena.

Hysterical overreactions to trivia are the signature of conflicts with high stakes. Benefits accruing to one side must be opposed, not because they are wrong or even significant, but simply because anything that strengthens the "enemy" is bad. If shortening shoelaces by a millimeter makes creationists happy, then we must immediately launch a national campaign to keep shoelaces unchanged or perhaps even made longer.

The cartoonlike character of the creation–evolution controversy was all too apparent in 1995, when the National Association of Biology Teachers (NABT) published its "Statement on Teaching Evolution." Nominally motivated by a desire to help high-school teachers navigate the troubled waters of evolution, the document was instead a rhetorical Trojan horse, designed to eliminate whatever tiny role students may have been retaining for God in the process of evolution. The central part of the document contained the following definition: "The diversity of life on earth is the outcome of evolution: an unsupervised, impersonal, unpredictable and natural process of temporal descent with genetic modification that is affected by natural selection, chance, historical contingencies and changing environments."[9]

The NABT supposedly wanted to be helpful in clarifying for high-school teachers just how evolution should be understood. But it's hard to imagine what the NABT was thinking, or if it was thinking at all. The definition is quite inadequate on its own terms and unnecessarily offensive to the very sensibilities that made teaching evolution complex in the first place.

For starters, nobody understands the trajectory of evolution well enough to make *unpredictability* a part of its *definition*. Evolution, as understood by some of its leading and most respected theorists, like Simon Conway Morris of Cambridge University, does have a direction. Taking direct aim at Gould, Morris suggests: "Rerun the tape of life as often as you like, and the end results will be much the same."[10] Conway Morris notes that many interesting properties of organisms, from compound eyes, to the ability of bats and certain birds to echo-locate, to the intricate social structure of ants and bees, have evolved more than once.[11] If evolution was entirely unpredictable, we would not expect this. Robert Wright, in his provocative book *Non-Zero: The Logic of Human Destiny,* argues from game theory that certain evolutionary trajectories are naturally favored over others.[12]

Capabilities like vision and intelligence are so valuable to organisms that many, if not most, biologists believe they would probably arise under any normal evolutionary process. I suspect that the majority of evolutionists, if informed that life had just begun evolving on some distant planet, would anticipate that vision and intelligence would eventually appear. So how can evolution be entirely random, if certain sophisticated end points are predictable? Evolution is like the path of a water molecule making its way down the side of mountain—unpredictable on a small scale but certainly not without a general direction. The NABT's claim that *unpredictability* should be a part of the definition of evolution was, to say the least, misleading.

The definition also states that natural selection, chance, historical contingencies, and changing environments are the factors affecting evolution. This list is presented as apparently exhaustive. How does the NABT *know* that these four factors are the only ones to be considered? Does it know already that no undiscovered laws of biochemistry and no mathematically preferred genetic patterns come into play?

The definition's greatest controversy arose from the words *unsupervised* and *impersonal.* These are peculiar terms in the context of a scientific definition. Since when is "supervision" something that science comments on? If the NABT read the definitions of other concepts in science, it would certainly have noticed that *nobody* uses the descriptor *unsupervised.* Do students learning chemistry or geology have to understand the natural phenomena of those disciplines as "unsupervised"? Similar problems attend the use of the word *impersonal.* The only possible role played by these two words is the expulsion of God from the evolutionary process. Who, exactly, is the "supervisor" who is not there? And what is the "personal" involvement being excluded?

Some distinguished philosophers pointed these problems out to the NABT, arguing that the definition made *theological* claims that went beyond science. Furthermore, they suggested that this definition would boomerang and ultimately prove counterproductive. Anti-evolutionists would eagerly endorse the definition, highlighting its clear incompatibility with religion, thus enlarging the gap between the scientific community and religious believers. The statement, they wrote, "gives aid and comfort to extremists in the religious right for whom it provides a legitimate target." Deleting the two loaded words would "defuse tensions" that were causing "unnecessary problems."[13] Wise counsel, indeed.

It would be nice to report that the NABT was simply careless in creating its inflammatory definition and, once that was pointed out, it happily changed it to reduce the controversy that makes evolution so hard to teach in the first place. However, this is not what happened.

The board of the NABT met in October 1997 to consider the recommendation that the theological terms be removed from the definition. After consideration, it voted *unanimously* to leave the terms in place and the definition unchanged. In pure culture-wars reasoning it explained that modifying the definition would give creationists "aid and comfort."[14] Never mind whether the terms were appropriate or not, the issue at stake was the comfort of the enemy. If doing the right thing comforts the enemy, then we mustn't do the right thing. It might be misconstrued as apologizing.

The NABT eventually made the suggested changes, but only after the level-headed and politically savvy anti-creationist Eugenie Scott convinced that body the offensive definition would come back to haunt it. The words *unsupervised* and *impersonal* were removed; the scientific content of the definition was, of course, unchanged by these deletions. Evolution was now like chemistry, geology, and football. God, if he exists, was allowed to watch.

WAR OF THE WORDS: ROUND TWO

A similar war of words occurred in 2001 when creationist senator Rick Santorum added the following language to an education bill dealing with the "No Child Left Behind" program:

> Good science education should prepare students to distinguish the data or testable theories of science from philosophical or religious claims that are made in the name of science; and … where biological *evolution* is taught, the curriculum should help students to understand why the subject generates so much continuing controversy, and should prepare the students to be informed participants in public discussions regarding the subject.

At face value, this paragraph looks quite sensible. After all, evolution is constantly in the news and students—and their parents—are certainly going to be interested in why. Evolution is the only topic in the curriculum that is there by court order. Once upon a time it was illegal to teach it in some states. The most famous intellectual contest on American soil was over evolution. Half the country rejects the theory, despite the confident endorsements of the scientific community. Would it not be prudent to help students understand why evolution is so controversial?

The Senate apparently thought so and passed the bill 91 to 8. The version passed by the House, however, did not contain Santorum's amendment, so a committee met to reconcile the two versions of the bill. By now, however, Santorum's amendment was out in the open, and culture warriors on both sides were talking strategy. Because creationists hailed the language as something of a victory, the champions of evolution became alarmed. Anything that makes creationists happy must be bad. The scientific community responded in the form of a letter not unlike the one that the NABT received earlier regarding evolution. The letter outlined objections to Santorum's amendment and, just in case the logic of the letter was not adequately convincing, it was signed by representatives of almost a hundred scientific organizations.

"The apparently innocuous statements in this resolution," they wrote, "mask an anti-evolution agenda that repeatedly has been rejected by the courts." They objected that the language "singles out biological evolution as a controversial subject," even though "from the standpoint of science there is no controversy." Evolution, they said, was like Einstein's theory of relativity—"robust, generally accepted, thoroughly tested and broadly applicable."[15]

Comparing evolution to relativity in this way is ludicrous, and I speak as someone who has taught both topics for years to college students. Relativity is a simple theory and easy to test in a comprehensive way; it deals with a limited range of phenomena and attempts nothing so ambitious as the reconstruction of the history of life on this planet. Virtually all of it was worked out decades ago, and so little remains to do that there is limited activity in the field. I can't recall the last time something significant emerged out of relativity theory, and one can teach the subject from a textbook that is fifty years old.[16]

In contrast, evolution is vibrant and challenging, with tremendous activity and daily breakthroughs. And although it is technically true that the scientific community is reasonably united behind evolutionary theory, there are significant controversies within the field about details. Two of its leading theorists, Dawkins and Gould, both penned massive works within the past few years defending very different explanations of how evolution works.[17] Conway Morris thinks they both got it wrong.[18] Evolution contains plenty of controversy. The combatants agree that evolution is true, but that is not the same as agreeing on how it occurred. But we don't want students to know this, of course, lest it make them vulnerable to creationism. Never mind that the controversy about how evolution works is the single most interesting topic in all of science.

In marked contrast to evolution, opposition to relativity is not constantly—or even occasionally—in the news. No law has ever ruled that it could not be taught in school, or that a competing view must have equal time, or that it, and it alone, must be taught. Most students don't even encounter relativity, since it appears only in advanced physics courses avoided by all but the most elite students. Virtually none of the senators receiving the letter assuring them that evolution was like relativity could have made even an introductory comment about relativity.

There is only one theory in all of science that generates constant controversy. Acknowledging that fact is hardly "singling out" that theory for special consideration. Evolution is already getting plenty of consideration. If a theory generates this much controversy, would it not be appropriate to take note of this in those classes where it comes up?

Despite the obvious problems with the letter, it was signed by the heads of every imaginable scientific society—and some unimaginable ones—most with no vested interest of any sort in the teaching of evolution. The American Astronomical Society signed it, as did its counterparts in chemistry, physics, geology, meteorology, and even mathematics and linguistics. Several psychological societies weighed in, as did the American Political Science Association. Geographers were represented—we can't have the creationists redrawing the coastlines or doing away with latitude. The president of the Freshwater Mollusk Conservation Society signed it. And we must not forget the American Fern or Clay Minerals societies. Even the CEO of "Shape Up America!" signed it.

What was going on here? Why was the president of the Freshwater Mollusk Conservation Society weighing in on this issue? What was at stake? Why were so many mighty soldiers being recruited to fight so tiny a battle?

On the other hand, we must wonder why the creationists were loading this sort of baggage onto an otherwise straightforward education bill, hidden at the back where it was unlikely to be seen. Phillip Johnson, the leader of the ID movement, had actually drafted the language for Santorum, so there clearly was a "conspiracy" of some sort to get this language into the bill. Obviously the creationists and ID supporters believed this bill would crack open some door through which they might smuggle something of interest to them into America's public schools.

The answer is, quite simply, that evolution has become the focal point of a culture war, which means that the goal of the protagonists is to *win*, not to discover the truth. Conceding minor points to your opponents, using inoffensive language, working out compromises, and finding middle ground are simply not allowed. Too much is at stake for such wimpy pussyfooting.

How else can we explain the offensive definition of the NABT? Or Santorum's sneaky insertion of language into an education bill? Or the crazed overreaction to the Santorum amendment, which, by the way, was removed, allowing the president of the Freshwater Mollusk Conservation Society to sleep much better at night.

EVOLUTION'S PERENNIAL CULTURE WAR

When Galileo quarreled with the Roman Catholic Church in the seventeenth century over the motion of the earth, the political dimensions of the conflict were lopsided. The church had the power to put Galileo on trial, sentence him to house arrest, and, if we can trust Catholic theology, excommunicate him and consign his soul to hell; in contrast, Galileo had little more than the strength of his arguments. And, at the time, these were not compelling.

Science in Galileo's century was young. It was both nurtured and constrained by the church. Because later developments vindicated Galileo, the verdict of history—at least popular history—is that the church abused its power in dealing with Galileo. This verdict has haunted the Roman Catholic Church ever since, as every generation has created for itself a new Galileo—a scientific martyr to wave in the face of the church when they disagreed with it.[19]

Two hundred years later, when Darwin published his controversial theory, science was a more substantial cultural force. Nevertheless, the church remained socially and politically powerful, and Darwin had to struggle against far more than simply opposing ideas. On an intensely personal level, this included his wife's theology and strong Christian faith. Darwin's persistent nervous disorders may have resulted from his concern about placing himself at odds with those he loved.

On a larger scale, Victorian society still retained many of the political structures through which the church had historically wielded its power. Ecclesiastical authorities sometimes maligned the champions of evolution for undercutting religion. Social pressures were brought to bear in ways that were deeply resented by honest scientists who simply wanted a fair hearing for their ideas. Like parents who can send children to bed when an argument starts going poorly, religious authorities were resented when they used the power of religion to settle disputes about which they knew nothing. Darwin's clerical contemporaries turned many away from evolution by claiming it was incompatible with the Bible. More recently, on the other side of the Atlantic, Henry Morris convinced millions of Americans that evolution is a satanic theory, at odds with the Bible and the Christian faith. As a political strategy this works wonderfully. No need to engage the scientific issues and open that can of worms—you simply poison the theory, so people will reject it without troubling themselves over whether it is right or wrong.

For all of its history, as we have seen in earlier chapters, evolution has been embedded in larger and often more substantial agendas than simply the history of life on this planet. William Jennings Bryan blamed it for World War I. Hitler's henchmen appealed to it to rationalize their genocide. Andrew Carnegie invoked it to promote unfettered capitalism. Eugenicists used it to justify mandatory sterilization of the "feebleminded."

Hostile creationists continue to blame evolution for everything from pornography to drug abuse. Evolution, in their eyes, is the root of all evil. Likewise, eager evolutionists appeal to the theory to explain rape and infanticide. Bystanders can't help but be nervous, but they dare not disbelieve, lest they turn into buffoons.

Those who oppose evolution, for whatever reason, or suggest that it might not be the full story, or look for some small role for God inevitably find ridicule raining down on their poor benighted heads. Dawkins, the leading public spokesperson for evolution, labels them "ignorant, stupid, or insane (or wicked ...).''[20] You *must* believe ...

EVOLUTION AS RELIGION

The promotion of evolution—both biological and cosmic—by its champions grows ever more evangelical as time goes by. Proponents sound more and more like preachers. Who can forget the priestly image of Carl Sagan standing behind his scientific pulpit on *Cosmos,* with majestic music and inspiring images in the background? The book titles sound increasingly religious—*Darwin's Cathedral, River Out of Eden, The Devil's Chaplain, The Creation, The Demon-Haunted World, The Dragons of Eden, The First Three Minutes, The God Gene, In the Beginning.*

Recently the long arm of evolutionary explanation has reached directly into territory where traditionally religious phenomena reside. Evolutionary theory now provides naturalistic explanations for altruism, morality, our religiosity and predisposition to believe in God, even the love we feel for our children. It explains "sin" and offers explanations for rape, infanticide, and the pervasive genocide that plagues our planet. Evolution now provides a rich and satisfying creation story—a scientific myth displacing the religious origins myth in the Bible. Evolution offers a source of meaning and an explanation for good and evil.

Is it any wonder that evolution and creation are locked in mortal combat? No longer do we seek a peaceful coexistence for science and religion, for the former now insists it has devoured the latter. The creation myth of our time is, as the original NABT definition stated, impersonal and unsupervised.

The obvious objection to all this is, of course, that it isn't true. Few evolutionary biologists think this way, and many are on record arguing that science has no business setting up camp in religious territory. Unfortunately, those who promote this conciliatory arrangement are a silent majority, all but invisible, missing from bookstands and public television.

In contrast, virtually all the leading spokespersons for science—the ones on bookstands and public television—are strongly antireligious. Even though religious belief is common in the scientific community, it is almost nonexistent among scientists who have become public figures. Richard Dawkins, Steven Weinberg, E. O. Wilson, Stephen Jay Gould, Carl Sagan,

Stephen Hawking, Steven Pinker, Francis Crick, Peter Atkins—all have (or had) considerable stature in the scientific community. But they are all hostile to religion and see it as something to be "explained away" by science. Even Gould and Wilson, the diplomats of the group, treat religion in a way that offends most religious people.

These are the scientists who have been setting the agenda, leading the larger cultural discussions of our time, creating the image of science in popular culture. They exert enormous influence on public perceptions of science and play roles in our society similar to those of the oracles of ancient Greece—delivering deep messages about the way things really are.

Dawkins, for example, is probably the leading public intellectual in the English-speaking world and uniquely a member of both scientific and literary societies; Wilson was recognized by *Time* as the seventeenth most influential person of the twentieth century and has won two Pulitzer Prizes; Weinberg and Crick are Nobel laureates. Gould has appeared on *The Simpsons,* in a special episode parodying his suggestion that science and religion should be "nonoverlapping magisteria." (The episode closes with a judge ordering religion to stay five hundred yards away from science at all times.) Hawking packs large auditoriums in his public appearances and has written a runaway best seller. Sagan was once one of the most recognizable people on the planet. Pinker is Harvard's "celebrity professor."

These thinkers—who all endorse fully naturalistic evolution with enthusiasm—are communicators par excellence. Their writings are models of clarity and eloquence; there is no doubt that they take communication seriously, despite being (or having been) active scholars within their fields who also publish in technical journals. Now, if all that Dawkins and company were doing was popularizing science, there would no cause for alarm. But an examination of the writings of this group reveals a larger agenda, one of breathtaking scope and ambition. In addition to lucid expositions of a wide range of scientific concepts from DNA to consciousness, we find suggestions that science should replace religion.

The idea that science should be a religion on its own runs like a subterranean reservoir through the writing of these popularizers, gurgling beneath the surface and bubbling into view every time the conversation gets to the now-here-is-what-it-all-means phase. In the closing paragraphs of long books about science, the exposition suddenly morphs into theology. The scientist is transformed into an oracle, telling us something grand and important that is, surprisingly, so much larger than the story unfolded in the previous pages. Readers are subtly carried to the top of a grand scientific mountain and offered a view of the promised land.

Marvel at the "ancestor's tale," writes Dawkins in a book of the same name, and note how much grander it is than the fairy tales of Genesis. Let science lift you above farce, says Weinberg, and provide some meaning in this pointless universe. Worship the evolutionary epic preaches Wilson. Harness the energy being squandered in traditional religions and redirect it where it might do some good; seek out a theory of everything, says Hawking, for there you find the mind of God. Celebrate the cosmos, says Sagan, for it is "all that is, ever was, or ever will be." Marvel at the luck that brought you here, says Gould, for natural history reveals no purposeful trajectory from simple organisms to us.

When the NABT proposed the controversial definition of evolution discussed above, there was at least a semblance of objectivity and, in the final analysis, it did provide a definition that was certainly less overtly antireligious than it might have been. The problem is that few people, except for high-school biology teachers and scholars following the creation–evolution controversy, have even heard of the NABT. Ordinary Americans are far more likely to encounter discussions of evolution, and science in general, in popular presentations.

Books about evolution, for example, appear on the nonfiction best-seller lists; PBS science programs and radio talk shows often deal with evolution; and leading evolutionary thinkers are often quoted in news stories. In such settings it is rarely advantageous to speak with dispassionate scientific objectivity. Audiences want excitement, hyperbole, and controversy; if you can provide that, you will be quoted. If you say that creationists are "stupid, wicked, or insane," journalists will return to you for commentary on subsequent controversies. The media are no respecters of scientific boundaries, and few journalists will scold a scientist for stepping outside the bounds of science to say something colorful, no matter how irresponsible.

When scientists speak as scientists, as they would when writing for scientific journals or presenting results at conferences, they are scrupulously careful to the point of tedium to maintain a strict silence on questions outside of science. Evolution, when discussed in the prestigious journals

Nature and *Science,* for example, would never be described as "unsupervised" or proposed as a replacement creation story. Critics would never be labeled "wicked" or "insane." But when a scientist writes or speaks to popular audiences, the rules change dramatically.

Dawkins is the worst offender. He has written many popular books on evolution, and in 2004 he published a 614-page opus titled *The Ancestor's Tale.* This magisterial work traces the history of life on this planet from its beginnings to the present, explicating our best understanding of that process. The content is mainstream science popularization carried along by outstanding prose and unencumbered by philosophical and theological asides. In the final three paragraphs of the book, however, as he draws his grand narrative to a close, he reflects on the meaning of what he has done:

> I have not had occasion here to mention my impatience with traditional piety, and my disdain for reverence where the object is anything supernatural. But I make no secret of them. It is not because I wish to limit or circumscribe reverence; not because I want to reduce or downgrade the true reverence with which we are moved to celebrate the universe, once we understand it properly. "On the contrary" would be an understatement. My objection to supernatural beliefs is precisely that they miserably fail to do justice to the sublime grandeur of the real world. They represent a narrowing-down from reality, an impoverishment of what the real world has to offer.[21]

Carl Sagan offered similar reflections in the final paragraph of the 345-page *Cosmos,* on which the television series was based:

> We are the local embodiment of a Cosmos grown to self-awareness. We have begun to contemplate our origins: starstuff pondering the stars; organized assemblages of ten billion billion billion atoms considering the evolution of atoms; tracing the long journey by which, here at least, consciousness arose. Our loyalties are to the species and the planet. *We* speak for Earth. Our obligation to survive is owed not just to ourselves but also to that Cosmos, ancient and vast, from which we spring.[22]

Stephen Jay Gould, who desperately wanted to be a mediator between science and religion, was nevertheless insistent that evolution was purposeless and without direction. Unchecked by the referees that brought the NABT to its senses, he wrote in the final paragraph of the 323-page *Wonderful Life:*

> And so, if you wish to ask the question of the ages—why do humans exist?—a major part of the answer, touching those aspects of the issue that science can treat at all, must be: because *Pikaia* survived the Burgess decimation. This response does not cite a single law of nature; it embodies no statement about predictable evolutionary pathways, no calculation of probabilities based on general rules of anatomy or ecology. The survival of *Pikaia* was a contingency of "just history." I do not think that any "higher" answer can be given, and I cannot imagine that any resolution could be more fascinating. We are the offspring of history, and must establish our own paths in this most diverse and interesting of conceivable universes—one indifferent to our suffering, and therefore offering us maximal freedom to thrive, or to fail, in our own chosen way.[23]

E. O. Wilson is arguably our greatest living scientist, the founder of the field of evolutionary psychology and humanity's most eloquent conservationist. His recent book *The Creation* was written as a series of letters to pastors, encouraging them to join him on his crusade to protect the environment. Nevertheless, despite the value he places on religious communities as partners in care for the planet, he ultimately intends that religion will be explained by science. Near the end of his Pulitzer Prize–winning *On Human Nature,* he writes:

> If religion, including the dogmatic secular ideologies, can be systematically analyzed and explained as a product of the brain's evolution, its power as an external source of morality will be gone forever and the solution of the second dilemma will have become a practical necessity.... What I am suggesting, in the end, is that the evolutionary epic is probably the best myth we will ever have. It can be adjusted until it comes as close to truth as the human mind is constructed to judge the truth. And if that is the case, the mythopoeic requirements of the mind must somehow be met by scientific materialism so as to reinvest our superb energies.[24]

Nobel laureate Steven Weinberg ends his classic *The First Three Minutes* with these widely discussed reflections:

It is very hard to realize that this all is just a tiny part of an overwhelmingly hostile universe. It is even harder to realize that this present universe has evolved from an unspeakably unfamiliar early condition, and faces a future extinction of endless cold or intolerable heat. The more the universe seems comprehensible, the more it also seems pointless.

But if there is no solace in the fruits of our research, there is at least some consolation in the research itself. Men and women are not content to comfort themselves with tales of gods and giants, or to confine their thoughts to the daily affairs of life; they also build telescopes and satellites and accelerators, and sit at their desks for endless hours working out the meaning of the data they gather. The effort to understand the universe is one of the very few things that lifts human life a little above the level of farce, and gives it some of the grace of tragedy.[25]

Identical sentiments can be found in the writings of other leading science popularizers as well. The few mentioned above are simply the best known and most influential.

For better or worse, mainly worse, the content and significance of evolutionary theory is communicated to broad audiences by people like Dawkins. Suppose you wander into a typical bookstore, say Barnes & Noble, and ask the manager for a book that would help you "understand evolution, what it is, and what it all means." The manager may likely point you to Dawkins's *The Blind Watchmaker*. A blurb from the *Economist* on the back cover suggests that the book is "as readable and vigorous a defense of Darwinism as has been published since 1859." E. O. Wilson of Harvard calls it "the best general account of evolution I have read in recent years." The author, a chaired professor at Oxford University, is well credentialed. Nowhere is there so much as a *hint* that *The Blind Watchmaker* is anything other than a superb articulation of Darwin's theory of evolution.

As you read the book, perhaps wondering about the relationship between evolution and your belief that God created everything, or at least was involved in some way, you gradually discover that evolution is absolutely incompatible with the idea that God created the world. *The Blind Watchmaker* presents you with a choice—either accept evolution and be on the side of science, enlightenment, progress, and truth or accept creation and be against science, on the side of superstition, darkness, and irrelevance. Just six pages into the book you encounter the following claim: "Although atheism might have been *logically* tenable before Darwin, Darwin made it possible to be an intellectually fulfilled atheist."[26]

Dawkins's claims might disturb you enough to make you seek a second opinion. Returning to the bookstore, you pick up other popular books by leading biologists and philosophers of biology and discover they agree with Dawkins. The leading spokespersons for evolution almost all say the theory refutes and replaces the traditional belief that God created everything. And many of them write with unbridled glee about this state of affairs, as if replacing belief in creation is the most important feature of evolution.

Now suppose you go to a Christian bookstore looking for another perspective. In a surprising twist, you encounter this identical argument in the writings of the anti-evolutionists. The architect of modern creationism, the late Henry Morris, describes evolution as Satan's "long war against God." And, despite the oddness of that claim, you take note that Morris is a solid academic, with a real Ph.D. and an impressive academic career. He is no fraud posturing with a fake degree and pretending to be a scientist. Morris and his fellow creationists believe that evolution has no supporting evidence and is literally nothing more than an alternative creation story to make atheists happy and fulfilled.

"There is no scientific proof," Morris writes, "that vertically-upward evolution occurs today, has even occurred in the past, or is even possible at all, yet it is widely promoted as a proven fact."[27] Nevertheless, despite evolution being scientifically vacuous, he concludes that "evolutionism is the proximate cause of the world's evils, for it is the basic belief and deceptive tool of Satan."[28]

Creationist Jonathan Sarfati, with a Ph.D. in physical chemistry, makes a similar point at the beginning of *Refuting Evolution:* "The framework behind the evolutionists' interpretation is naturalism—it is assumed that things made themselves, that no divine intervention has happened, and that God has not revealed to us knowledge about the past." "Evolution" Sarfati argues, "is a deduction from this assumption" rather than an inference from observations of the natural world.[29]

Phillip Johnson's strategy is based on his conviction that a blinkered and deluded commitment to naturalism is the reason scientists can't see the weaknesses of evolutionary theory—why they miss the clear evidence for design in DNA, in the blood-clotting mechanism, in the flagella of the bacte-

ria, and elsewhere. In his assault on naturalism, titled *Reason in the Balance,*
he writes:

> What is presented to the public as scientific knowledge about evolutionary
> mechanisms is mostly philosophical speculation and is not even consistent
> with the evidence once the naturalistic spectacles are removed. If that leaves
> us without a known mechanism of biological creation, so be it: it is better to
> admit ignorance than to have confidence in an explanation that is not true.[30]

For those whose worldviews include Satan as an omnipresent evil per-
sonality, evolution is precisely the sort of deception that makes sense. Convince people they are the product of a random, purposeless, cruel process,
and atheism, moral anarchy, and decadent reality shows about wife swap-
ping won't be far behind. Even without Satan the charges above are damning, although not quite so literally. If evolution starts with the assumption
that there is no God and then selectively and deceptively assembles a case to
rationalize this starting point, the scientific community is no different than a
team of creepy defense lawyers working to free rapists and serial killers they
know are guilty.

Such ad hominem attacks on the integrity of science seem unfair. Surely
the scientific community that put a man on the moon, wiped out smallpox,
and built the iPod is not engaged in this sort of shady enterprise. Significantly, though, some prominent members of the scientific community agree
that their enterprise *is* all about making God obsolete. They admit they will
defend and even promote preposterous notions, rather than admit that God
might have some relevance to understanding the natural world. A leading
geneticist, Richard Lewontin, has stated this commitment with impressive, if
unrepresentative, candor:

> We take the side of science *in spite* of the patent absurdity of some of its
> constructs, *in spite* of its failure to fulfill many of its extravagant promises
> of health and life, *in spite* of the tolerance of the scientific community for
> unsubstantiated just-so stories, because we have a prior commitment, a com-
> mitment to materialism. It is not that the methods and institutions of sci-
> ence somehow compel us to accept a material explanation of the phenomenal
> world, but, on the contrary, that we are forced by our *a priori* adherence to

material causes to create an apparatus of investigation and a set of concepts that produce material explanations, no matter how counter-intuitive, no matter how mystifying to the uninitiated. Moreover, that materialism is an absolute, for we cannot allow a Divine Foot in the door.[31]

Lewontin's honesty is interesting and, despite what looks like a disturbing admission of the very blindness that Johnson has assaulted, Lewontin's views are echoed by many scientists who have taken the time to describe its inner workings.

Another evolutionary biologist describes science as a "game with one overriding and defining rule," namely, that science "explain the behavior of the physical and material universe in terms of purely physical and material causes, without invoking the supernatural."[32] A leading Cornell University historian of science, William Provine, has written that "biology leads to a wholly mechanistic view of life." This view cannot be reconciled with belief in God: "The frequently made assertion that modern biology and assumptions of the Judeo-Christian tradition are fully compatible is false."[33]

THE CULTURE WAR

Viewed by these lights, the creation–evolution controversy is far more than a debate over the origin and development of life on this planet, the age of the earth, or the relationship between humans and the rest of the animal kingdom. The controversy is about the larger question of who decides what the nature of ultimate reality is. Will Dawkins and his merry band of materialistic naysayers provide the creation story for our culture? Or will it be Johnson and his underdog team of designer Christians? Dawkins and Johnson agree that the choice is a real one, the alternatives are incompatible, and the consequences significant.

In his recent projects, Dawkins clarifies that his agenda is not simply the promotion of evolution, or even the "public understanding of science," as his endowed professorship at Oxford is titled. His agenda is the destruction of religion. He produced a documentary on religion for British television titled "The Root of All Evil." His recent book was called, provocatively, *The God Delusion*. His 2003 Tanner Lectures at Harvard were titled *The Science of Religion and the Religion of Science*. On these occasions he assaulted

religion with venom not seen since an angry mob offered Jesus up for cru-
cifixion. Dawkins and his followers, in their Oxbridge and Ivy League pro-
fessorial robes, lament that the great engine of secularization has stalled and
religion is making a comeback. Forget Darwin and widespread cultural confusion about evolution; the troops must be rallied to oppose religion.

Johnson is a mirror image of Dawkins. He sees in naturalism the same
pernicious cultural cancer Dawkins sees in religion. His interest in evolution derives entirely from its role as an important part of the foundation for
naturalism. His agenda for destroying naturalism is to use weaknesses in
evolution as openings into which to insert his "wedge." Everyone knows,
however, that if evolution collapsed without also bringing down naturalism,
he would keep on fighting.

Dawkins, Gould, Weinberg, Provine, Pinker, Dennett, and Atkins versus
Johnson, Morris, Dembski, Wise, Wells, Meyer, and Behe. Atheism versus
theism. Evolution versus creation.

Evolution has been embroiled in this kind of controversy since before
Darwin published *On the Origin of Species by Means of Natural Selection*
in 1859. Enthusiastic polemicists were forever looking into the deep well of
its grand story and seeing their reflections. Thomas Huxley saw a weapon
to wield against the clerics and their archaic political power. Andrew Carnegie saw a rationalization for unbridled capitalism. Privileged Victorians saw
a rationale for ignoring the plight of England's poor. Herbert Spencer saw
a "might makes right" moral code. William Jennings Bryan saw the roots of
World War I. Nazis saw a rationale for genocide. American social planners
saw a rationale for eugenics. Fascists looked at evolution and saw fascism;
Marxists saw Marxism; free-market enthusiasts saw capitalism. Now today,
Harvard's Steven Pinker sees in evolution an explanation for infanticide;
Randy Thornhill and Craig Palmer see an explanation for rape.[34] Evolutionary psychologists see the genetic basis of male philandering and playground
bullying. But they also see the genetic basis for brotherly love and sacrificial
care for one's children.

Is this Darwin's theory of evolution—this catchall story of origins that
can be adapted as the scientific basis of everything from capitalism to
brotherly love? For the majority of scientists, excluding the few who write
books with titles like *The God Delusion,* evolution is simply the central idea
in biology. They would like to see evolution taught in America's high-school
biology classes and are frustrated that this poses such a problem.

Darwinism, however, cannot escape its rich, complex, troubling, exhilarating, sobering, and inspiring history. Evolution in the labs, in the field, and in the textbooks may actually be nothing more than a central biological theory of great utility, uniting a broad range of natural phenomena under a single explanatory umbrella. But, in ways that have no analog anywhere else in all of science, evolution is connected to a host of other ideas, some very disturbing. To suppose, as so many do, that evolution can be disconnected from these ideas and taught purely as science is naive. To argue, on the other hand, that evolution derives from these ideas is simply wrong. No wonder the conversation is going nowhere.

CHAPTER 8

EVOLUTION AND PHYSICS ENVY

O n September 22, 1919, Albert Einstein received a telegram that said: "Eddington found star displacement at rim of sun." Sir Arthur Eddington was England's greatest astronomer. And he had just determined that light beams from stars in the Hyades cluster deflected as they passed near the sun on their way to earth. The deflection made the stars appear in a different location in the sky, just as a spoon in water will appear to be bent. Eddington's ambitious observation, made during a total eclipse on an island off the coast of West Africa, tested Einstein's theory of general relativity. This theory, destined to overturn Isaac Newton's venerable explanation for gravity, suggested that "empty" space was *warped* by gravitational masses like the sun. As a test of his novel theory, Einstein predicted that light beams passing through this warped space would be deflected.

Einstein's prediction was bold and reckless. If the light had not deflected, his theory would have collapsed and a decade of hard work would have been lost. But when his prediction came true, the theory was confirmed. An oversized *New York Times* headline on November 10 declared, "Lights All Askew in the Heavens," celebrating the arrival of Einstein's theory. The few who understood were deeply impressed. A theory challenging Newton's durable explanation had predicted an exotic physical effect. Eddington observed the effect. Newton, quite literally, had just been eclipsed by Einstein.

The precision and rigor of these tests of Einstein's theory impressed Sir Karl Popper, Europe's greatest philosopher of science. How could a theory so remarkably confirmed *not* be true? And what intellectual courage and confidence Einstein showed in developing such a *risky* prediction—a prediction that could have *falsified* his theory. Surely this was science at its best—

rigorously empirical, testable, objective, and, ultimately, *true*. Should not the generation of such predictions be the hallmark of *all* scientific theories?

Popper developed this idea into an influential definition of science. All genuinely scientific theories, he argued, must make novel predictions about unknown phenomena. These predictions must be articulated so clearly that that they can be conclusively refuted by observation. And if the predictions fail, the theory has been falsified. If a theory *cannot* make such falsifiable predictions, then it cannot claim to be scientific.

Evolutionary theory, however, because of its scope, complexity, and dependence on history, does not lend itself to this kind of simple analysis, which led Popper to reject it initially as a pseudoscience. The scope of evolution's explanatory power is breathtaking, however, and eventually Popper changed his mind. In this chapter I want to spotlight evolutionary theory's remarkable capacity to unite disparate observations of the natural world. Seemingly unrelated patterns in nature become part of a coordinated package when brought under the explanatory umbrella of evolution, although not exactly in the simple and elegant way that Popper would have preferred.

Despite its problems, the falsifiability criterion developed by Popper was broadly admired and promoted as a simple test to distinguish authentically scientific ideas from pseudoscientific imposters. As the twentieth century unfolded, the falsifiability yardstick would be laid alongside many ideas to see how they measured up. In the 1982 creationism trial in Little Rock, Arkansas, for example, Judge William Overton ruled that creationism could not make falsifiable predictions and thus was not science.[1] The creationist claim that an invisible being using processes not now operating created all life on earth could not generate falsifiable predictions. Therefore, it could not claim to be scientific.

Creationists, in a clever response, use Popper's falsifiability criterion to argue that *evolution* is not science.[2] In his autobiography, *Unended Quest*, Popper fanned this particular flame by lumping evolution together with Marxism and Freudian psychology as pseudoscientific metaphysics. He later changed his mind about evolution,[3] a reversal that has escaped the attention of the creationists, who continue to invoke him.[4]

Popper contrasted these pseudosciences with general relativity, noting that the latter made truly falsifiable and thus genuinely scientific claims about the world. The vagueness and unlimited flexibility of Freudian,

Marxist, and, for a time, Darwinian explanations distressed Popper. Marxism, to take one example, predicted the exploitation of workers by bosses, but when counterexamples were found—such as companies that paid good salaries to their workers—Marxists would explain those as a different or subtler form of exploitation. The unexpected counterexamples were strangely incapable of falsifying the theory. Marxism, in fact, couldn't seem to make *any* predictions that, if they failed, would refute the theory.

Creationists—and the early Popper—see the same sort of wishy-washy can't-be-falsified explanations in evolution. Early Darwinists, and Darwin himself, predicted the existence of countless transitional forms in the fossil record. When those were not found, however, the theory was adjusted to accommodate the failed prediction rather than rejected as falsified. This contrasted dramatically with general relativity, which handed an ax to Eddington and then placed its scientific neck firmly on the chopping block of observation.

Pseudosciences, argued Popper, were not science, but *ideology*, and their signature was the blind devotion of their advocates, who forever adjusted their "theory" to square it with anomalous data, rather than subject it to genuine testing. Eventually such false ideologies masquerading as science would be exposed and abandoned.

Popper began his thoughtful and devastating critiques of the theories of Freud, Marx, and Darwin in the early decades of the twentieth century. Subsequent developments proved him partially right. Marxism has indeed all but died except around a small table in North Korea; its central ideas turned out to be anachronistic ideology, and its founder has faded into the canvas of history. Ditto for Freud. Darwinism, in contrast, has not died, but rather has grown steadily stronger and more influential. Few philosophers today would reject it as unscientific. And, of course, even Popper, in a rare act of intellectual humility, reversed his earlier stance on evolution.

Nearly a century later, relativity and evolution are still being juxtaposed. When the National Association of Biology Teachers defended evolution against charges that it was "only a theory," it compared it to relativity. If relativity's claims to truth were not compromised by its status as a theory, then evolution, it argued, can hardly be criticized for being "only" a theory.

But, on the other hand, creationists and intelligent design enthusiasts remind us just how inferior the theory of evolution is to theories in physics. Physical theories present their conclusions in tidy mathematical equations—

think $E = mc^2$. The relevant phenomena can be demonstrated in laboratory experiments and in public displays at science museums. Impressive technological spin-offs bathe the underlying science in the warm glow of credibility. Evolution, alas, offers nothing but vague generalities—"the fittest survive"—and invokes entities like "common ancestors" or processes like "speciation," for which the evidence is often depressingly small and indirect. That slippery and mysterious character named "chance" plays a central, but vague, role in the great drama of evolution. The process is driven by an elusive and all-but-unobservable metaphor called "natural selection" conferring "reproductive advantage" on organisms. Most of the work of evolution is done by mutations that occur in species that go extinct and leave no trace. Species that go extinct without leaving any evidence that they ever existed bear an unfortunate resemblance to fairies and leprechauns.

The theory of evolution, embedded in biology as it is, bears little resemblance to theories in physics. The disciplines are quite different. The phenomena they study have little overlap, and even the scientists in the two fields are different. When Francis Crick, who won the 1962 Nobel Prize for determining the structure of DNA, moved from physics to biology, he found the transition so dramatic that it was "almost as if one had to be born again." Crick recalls having to consciously abandon the physicists' intuition about nature's "elegance and deep simplicity." Physicists, he warned, are apt to "concoct theoretical models that are too neat, too powerful, and too clean."[5]

Crick observed what Popper had noted a few decades earlier. But Popper, like most philosophers of science before and since, had been bewitched by the grandeur of physics, with its elegant laws and imposing mathematical language. He saw it as the paradigm for all of science, setting an unreasonable standard that other sciences could not possibly reach.

THE SIMPLE LIFE

Physics is science at one extreme of simplicity. Physicists study incredibly simple natural phenomena, like the forces between bodies in space or the behavior of electrons orbiting around nuclei—phenomena that can actually be *thoroughly* understood. I earned a Ph.D. by studying helium atoms for three years and understood them very well by the time I graduated. By restricting its focus to simple systems, physics produces seductively elegant explanations. These explanations are expressed in compact mathematical equations making specific quantitative predictions. The simplicity and thoroughness of such explanations, however, are not due to physicists' superior scientific practices, as many have misconstrued, but to physicists' selection of the simplest problems on which to work. It is no accident that physics was the first science to develop historically or that its first major accomplishment—showing that the earth went around the sun rather than vice versa—was both theoretically trivial and mathematically elegant.

Creationists and ID enthusiasts like to argue, in concert with the early Popper, that evolution is not a science because it is not based on rigorous empirical evidence like physics; there are no "evolution in action" shows at the science museum to go with the whiz-bang electricity demonstrations. In contrast, the NABT argued exactly the opposite—that evolution is a science because it *is* like physics. Both arguments are hopelessly flawed.

THE STORY OF EVOLUTION

Evolution is a solid and robust scientific theory, because it explains many things about the world and relates countless otherwise disconnected facts to each other. It is *not* a science because it resembles physics. Evolution is a messy theory, however, with a history of dumb mistakes, serious errors, occasional fraud, and overconfident assertions. When its problems are gathered and packaged by clever polemicists like Phillip Johnson, Ken Ham, or the late Henry Morris, evolution comes off looking rather pale. Such a judgment, however, is uninformed. To be sure, the theory of evolution does indeed have problems, but these are little more than tiny holes in a vast tapestry of compelling explanation.

The fossil record, for starters, shows an unmistakable trajectory from simple to complex as we go from ancient strata to more recent. The distribution of animals around the globe, called biogeography, shows a clear pattern that suggests that closely related species evolved from each other. The universality of DNA as the structural language of every life-form suggests a global relatedness of all species. And the details of specific DNA patterns link different species to common ancestors with the same clarity that DNA evidence in modern courtrooms links criminals to their crimes. Mutations observed in species that are easy to study, like fruit flies, disclose a genetic code perched on the knife edge of predictability and creativity, exactly the kind of balance that enables reliable evolutionary change over time. Studies comparing the anatomy of different species reveal intriguing similarities that make no sense outside evolution. Developing embryos of different species show strange coincidences that make sense only if those species are related. New computer models based on evolution offer interesting explanations for such things as our preference to remember our cousins in our will rather than our neighbors.

These areas of investigation are quite independent of each other. The fossil data, for example, began to accumulate in the eighteenth century, two hundred years before DNA was understood and long before there was an evolutionary interpretation of that data. Biogeography was controversial before Darwin was born, as European naturalists argued with their American counterparts over which continent had the more important species.

The theory of evolution does not claim to be true because of a single dramatic prediction about the natural world, as was the case with relativity. No prediction made by evolution comes close to Einstein's prediction that gravity would bend light and cause stars in the heavens to appear in new locations. But it is precisely the whiz-bang character of relativity that makes it, in the final analysis, a narrow theory explaining a limited range of phenomena. A theory that can be dramatically confirmed by one observation can hardly explain a gigantic roster of disparate phenomena.

Evolution makes up for its lack of precision and mathematical rigor with its astonishing scope. Before Darwin, who could have imagined that the same theory would explain both the fossil record and the peculiar genetic similarities between disparate organisms? Who could have imagined that when genes and DNA were finally understood, they would confirm relationships between species that had already been inferred from other data, such as comparative anatomy? The convergence of so many unrelated lines of investigation is a compelling argument for the truth of evolution.

LIFE'S GRAND STORY

The modern story of evolution, as Richard Dawkins makes so clear, is a grand tale, evoking wonder and mystery. It crackles with surprise and controversy and raises deep questions. I want to outline briefly what evolution claims about the history of life on this planet, and then look at the lines of evidence suggesting that this story is, indeed, true. Evolution, although not without its puzzles and controversies, is now so well supported that it demands our assent. It also demands our rejection of the various alternatives at play in America.

The story begins with the appearance of the first living cell on the earth roughly four billion years ago. This singular event was preceded by ten billion years of cosmic evolution from the big bang, through the origins of the elements in stars, through the appearance of planet earth with its "just right" conditions for life, to the emergence of a chemical environment capable of hosting the first living cell. There is presently no generally accepted theory of how the first life-form arose, but several options have been proposed. The raw materials, of course, were not alive, but were capable of assembling into a complex structure with the capacity to reproduce itself. And once reproduction was initiated, evolution began.

The first cell made copies of itself. A primitive genetic code guided the cellular machinery to gather material from the local environment to enable this copying. Thus one cell became two, two became four, four became eight, and eight became many billion. Single-celled life was simple and robust. It was the only form of life on the planet for over three billion years and flourishes today in the form of the ubiquitous, resilient, and inextinguishable bacteria.

The machinery enabling a single cell to make another version of itself was an intriguing combination of accuracy and flexibility—accurate in that every copy was largely the same, but flexible in that there was room for small changes to occur without disrupting the entire process. Slightly different cells that could copy themselves faster or more often had an advantage. A cell reproducing itself slightly faster than its peers will, over the course of a million years, take over the world and drive its peers to extinction.

About a half billion years ago a change occurred that enabled single cells to clump together. Perhaps the chemical composition of the external membrane was altered so they could stick together, like primordial Velcro. In any event, multicellular life appeared, bringing with it the possibility of greater complexity. A collection of cells can specialize in ways that a single cell cannot. The cells on the outside can learn to monitor the external world and protect the interior cells from threats, like border guards in a country, vigilant about external invaders but undistracted by internal matters. The interior cells can redirect their resources toward other functions, like reproduction. And the whole can become greater than the sum of the parts as specialization takes over.

A complex, specialized, multicellular organism can evolve along many different paths. Reproductive flexibility can lead to exterior changes like hair, scales, or feathers. External light-sensitive cells can become sophisticated and turn into eyes. A central nervous system can become intelligent. Lungs and kidneys can clone backups. And so on. The more complex the organism, the more things there are to change and improve.

Reproductive flexibility means that virtually all of the members of a species will be slightly different from each other. These differences will include variations in just about everything. Variations enhancing the production of offspring will result in more organisms with those variations, until gradually every member of the species has them. Catholics, for example, believe in large families, which is why there are so many Catholics. Variations irrelevant to procreation will turn out to be, not surprisingly, irrelevant. And those that interfere with reproduction—like the Shaker sect's belief that sex is wrong—will gradually disappear.

New species spin off from their parent species, often because a geographical barrier, like a river, slices through their habitat. Separated from parent species and confronting different reproductive challenges, the orphaned group evolves along its own path until eventually it can no longer interbreed with the parent species of which it was once a part. At this point we say that a new species has appeared; the parent species remains the common ancestor of both this new group and any others that spin off.

This is the trajectory of life on this planet; genetic flexibility constantly tosses out novel variations to be challenged by Mother Nature. If they enhance reproduction, they persist and spread; if they don't, they diminish and disappear.

Sometimes dramatic events intrude and alter the normally imperceptible course of evolution. Seventy million years ago evidence indicates a huge asteroid struck the earth, creating a gigantic crater on the edge of the Yucatan Peninsula. The dinosaurs could not handle the accompanying atmospheric disturbance. It interfered with their reproduction, and they went extinct. The departure of the dinosaurs created space for mammals, which at the time were small and insignificant. Gradually they began to prosper and proliferate, and many new species appeared.

A few million years ago, on the rich terrain of Africa, one mammal species—an apelike primate—began to walk upright, and Mother Nature smiled on the innovation. The new bipedalism spread to other species; before long some of these bipeds began to make tools with their newly available "hands." In one species the brain increased dramatically in size and the capacity for speech appeared. And finally, maybe a hundred thousand years ago, humans appeared. From an evolutionary point of view, nothing much has happened since then.

This is the evolutionary story, as developed by thousands of scientists working in countless disciplines from genetics to geology over the past two centuries. Darwin, of course, is the intellectual father of the theory, but his work built on those who went before him and has been extended in significant ways by the countless scientists who came after him.

HAS THE JURY REACHED A VERDICT?

Critics of evolution claim that the story told above is a "just so" story—countless cute and largely imaginary anecdotes strung together to create a naturalistic account of origins. The story is without foundation, they charge, with nothing to commend it beyond its avoidance of supernatural explanations.

Johnson, using Popper, or at least his vocabulary, charges that evolutionary science is filled with "pseudoscientific practices" because the relevant scientists are simply too stupid to understand the difference between the "scientific method of inquiry" and the "philosophical program of scientific naturalism."[6] He claims that scientists' blinkered embrace of "dogmatic metaphysical naturalism" leads them to "disregard some aspect of reality that is virtually staring them in the face.[7] Duane Gish, the venerable creationist debater and critic of the fossil evidence for evolution, describes evolutionary theories as nothing more than "pointless speculation, totally devoid of empirical evidence."[8] The Moonie creationist Jonathan Wells has written an entire book arguing that the most celebrated evidences for evolution—the "icons"—are all "false or misleading" and that evolutionists often don't even know that many of their favored evidences have been conclusively refuted.[9]

What is striking about these unrestrained assaults on evolution is their assumption that evolutionary biologists are too stupid to understand the situation. These dumb biologists confuse philosophy and science; they don't know their own field; they can't see that evolution is their religion and their belief in it a faith; they can't follow a simple argument or identify a preconception. Biology, apparently, is a field filled with morons and knuckleheads.

This conflict becomes ludicrous when we consider the relative credentials of the critics and those of the scientists they are attacking. Johnson is a lawyer—a bright one, to be sure, but without training of any sort in science. He understands so little about science that he actually celebrates his scientific illiteracy as an *asset,* claiming that this issue needs lawyers with rhetorical skills, not scientists who understand biology.[10] Ken Ham, probably the most influential creationist as of this writing, was a high-school teacher before he became a full-time "creation evangelist." There is something bizarre about Ham, who writes books with titles like *D Is for Dinosaur,* and his sweeping criticisms of the entire enterprise of modern science: "Most scientists do not realize that it is the belief (or religion) of evolution that is the basis for the scientific models (the interpretations, or stories) used to attempt an explanation of the present."[11]

The claim that evolution has no facts supporting it is quite ridiculous. We can argue, to be sure, that the facts might be interpreted in some other way; but to claim that there are no such facts is absurd. Books written by those who make such claims should be read for nothing more than their entertainment value.

The theory of evolution is a vast and complicated network of interlocking explanatory concepts tying together everything from the age of fossil bones to similarities between human and chimp DNA. There is, quite simply, a *mountain* of evidence from multiple sources supporting evolution. Organized by evolutionary theory, this mountain of evidence becomes a comprehensible and manageable landscape. Without evolutionary theory, it disappears into the clouds, a hidden and impenetrable mystery of unexplained patterns.

In no less than five distinct areas, patterns have been discovered that point strongly toward evolution. The confidence that biologists have in evolution derives from the way these lines of evidence converge independently to yield the same explanation. The five lines of evidence, each of which we will look at briefly, are:

The fossil record

Biogeography

Comparative anatomy

Developmental similarities

Comparative biochemistry/physiology

THE FOSSIL RECORD

Compelling arguments that the fossil record supports evolution come from history. Nineteenth-century geologists—most of them believers in biblical creation, as we have seen—were forced by discoveries to modify their belief in a recent sudden creation.

It started when the shovels and pickaxes of the industrial revolution unearthed fossils of many extinct species. Initially they were thought to be the residue of Noah's great (and worldwide) flood, but it was soon clear that many of these animals had never coexisted with humans. Not a single human fossil, for example, was ever discovered with that of a dinosaur, suggesting that the dinosaurs must have belonged to a previous era. Challenging this interpretation, Ken Ham shows dinosaurs being marched into Noah's ark in his creation museum, one of the more colorful claims circulating in fundamentalist circles. If Ham is right, then Noah's flood would have destroyed vast numbers of both humans *and* dinosaurs. Why not even *one* of these unfortunate humans managed to get buried in the same strata as a dinosaur is a deep mystery, if they all drowned together.

Decades before Darwin suggested evolution, geologists recognized that the fossil record spoke clearly of a long natural history that preceded the appearance of humans. Committed to the biblical story of creation, these geologists found ways to reinterpret the Genesis story. Perhaps the "days" of creation were geological epochs; maybe there was an earlier creation before the one described in the Bible. None of these creative reinterpretations proved satisfactory, however, and eventually Darwin's theory provided a simpler explanation.

The argument from fossils is particularly compelling when we realize that much of the data was in hand *before* Darwin. This was not a case of data being gathered to support evolution, but rather of data that seemed mysterious and puzzling until evolution came along to explain it. By the end of the eighteenth century geologists had established that stratified rock—such as would be exposed along the sides of a trench dug to carry railroad tracks—contained fossils in a clearly sequential order. But what did this mean?

Stratified rock, we now understand, tells a simple story that lets scientists see into the past. The stratification of rock is like the layering of a cake. Suppose you make a layer cake and you lay down the base of the cake at 2:00. You put some frosting on this first layer at 2:10 and then add another cake layer at 2:20; you frost this layer at 2:30. At 3:00 you write "Happy Birthday" on the top with colored frosting. A slice cut from this sort of layer cake is like a slice cut into the earth, both of which expose a history. The cake slice reveals the history of the cake. The frosting in the middle is older than the frosting on the top. A mosquito embedded in the first cake layer died before the frosting was put on; parmesan cheese dropped from a pasta dish whisked by at 2:45 will be located on the top layer of frosting, but under the "Happy Birthday" greeting.

Stratified rock layers record history in the same way. The dinosaurs in layers under those containing human fossils are there because they died eons before humans appeared. A thin layer of meteoritic dust at the boundary where dinosaurs disappeared indicates that a huge extraterrestrial mass hit the earth at the time the dinosaurs became extinct. The presence of Neanderthal fossils in the same layers as human fossils indicates that they coexisted.

The history displayed in this stratification gradually became clear, as paleontologists discovered that the sequence of fossils revealed a trajectory from simple to complex. Early fossils were simpler than later ones. The durable stone tools fashioned exclusively by the higher primates, for example, are completely missing from older rock strata.

These patterns were not concocted to support evolution. They were a part of the confusing picture painted by nineteenth-century science that eventually strained the credibility of the traditional biblical story until it could no longer be stretched and twisted to accommodate the data. But all this happened *before* Darwin's theory appeared.

The patterns have proven to be remarkably consistent and are now found around the globe. There are exceptions, to be sure, and Duane Gish has them all catalogued in his creationist classic *Evolution: The Fossils Say No!* The exceptions, however, all come with their own explanations. Most missing fossils are missing because those species went extinct before they had a chance to fossilize. The out-of-order fossils occur when a great thrust or fold of the earth's crust occurs, for example, in an earthquake. Such departures from the norm, however, are as obvious as a layer cake run over by a bicycle and then packed by hand back into its original shape.

In the two centuries since these patterns were first discovered the overall sequence noted in the geological record has been amazingly consistent. The discovery and classification of fossils is now a highly active field, and we can only be impressed that Darwin's original explanation has proven to be such a reliable guide, constantly confirmed by new discoveries.

Once Darwin's theory became a serious candidate to explain the history of life, it attracted more attention as scientists pondered its implications. A puzzle arose. All life-forms supposedly originated from a single common ancestor that evolved gradually into other species. It follows that the history of life must have included every imaginable transitional form as one species evolved into another. But the evidence suggested otherwise.

At the time Darwin wrote *On the Origin of Species* in 1859, there were huge gaps in the fossil record. Darwin was both puzzled and bothered by the absence of certain intermediate forms. Major groups of organisms appeared suddenly in the fossil record, looking as if they had been inserted there from outside, as if God were randomly performing piecemeal acts of creation. Creationists, of course, found this state of affairs to their liking, proclaiming confidently that the missing transitional forms were not "missing," but rather had never existed.

Much of Darwin's misapprehension derived from a simple lack of data. Many of the gaps in the fossil record have been filled in by subsequent discovery, and a great many transitional forms that creationists confidently assured us did not exist have been discovered. To be sure, there are still gaps in the fossil record, but enthusiastically pointing them out has become a bit like crying "wolf."

Hundreds of thousands of fossils now demonstrate transitions from one life-form to another. Discovered in precisely dated rock samples, they have filled in many of the gaps that bothered Darwin. Intermediate forms are well established between fish and amphibians, between amphibians and reptiles, and between reptiles and mammals. These are the large-scale transitions. On a finer scale there are less dramatic transitional forms illuminating smaller changes. The human family tree is especially well documented since it is more recent. An impressive fossilized trajectory has been unearthed, showing the evolutionary pathway from *Homo ergaster* to *Homo mauritanicus,* to *Homo heidelbergensis,* and beyond to *Homo sapiens.* Human evolution is so clearly on display in the fossil record now that one paleontologist has called it "the creationists' worst nightmare."[12]

We now have a catalog of fossils beyond Darwin's wildest imagination. The story told by the fossil record is an ever more detailed version of the one that Darwin told a hundred and fifty years ago. The oldest forms of life on earth were microbial. Rocks three and half billion years old contain evidence of bacterial life. The oldest evidence of more advanced and complex eukaryotic cells come from two-billion-year-old rocks. In much younger strata are found multicellular organisms, after which we find plants, fungi, and animals, the latecomers. The genus *Homo* and the species *Homo sapiens* in this story have just arrived.

SPECIES DISTRIBUTION, OR BIOGEOGRAPHY

The second important large-scale pattern pointing to evolution comes from biogeography, which studies the geographical distribution of species. Life is distributed on our planet in a most curious way. Tiny Hawaii, for example, is home to fully one-quarter of the two thousand species of fruit fly. Australia has an odd collection of animals found nowhere else. Darwin's celebrated Galapagos Islands are home to many different species, most of which resemble species on the mainland six hundred miles away.

These patterns make no sense in the absence of evolution. In the creationist picture all these animals descended from the pairs that disembarked from Noah's ark a few thousand years ago in Turkey. Astonishingly rapid speciation at biologically impossible rates would be required to produce all the animals that populate the Galapagos in the short time since Noah's flood.

Bring in the long history of life, however, and things fall nicely into place. Evolution predicts that speciation occurs most naturally and rapidly in small populations that get cut off from their parent species. In the case of the fruit fly, it is likely that a small population of fruit flies made it to Hawaii ages ago and found their new habitat quite congenial. Newly arrived, they had no predators to worry about and many available niches to occupy. These are the conditions for rapid speciation, as small subgroups broke away and found comfortable new homes. On the different Galapagos Islands, which are actually the tops of submerged volcanoes, there are three different species of mockingbirds, each on its own island. Darwin inferred correctly that they evolved from the single parent species on the coast of South America, six hundred miles away. Ages ago some of these mockingbirds relocated to the Galapagos Islands, and the separate populations evolved independently into different species.[13] The similar long isolation of Australia accounts for its peculiar species.

Across the planet the same patterns of speciation are increasingly apparent and explained as the result of evolutionary history. Biogeography has repeatedly led to novel predictions that have been confirmed, such as the existence of North American camels. There are camels today in Asia and Africa; their close relatives, the llamas, are found in South America. If this linkage is due to evolution, there should be camels in North America, which clearly there aren't. This led to the prediction that there should be extinct species of camels in North America, which were eventually found.[14]

Examples like these may seem minor, but there are so many of them that, taken collectively, they strongly and clearly support evolution. In fact, evolution is supported by many small pillars such as these.

COMMON STRUCTURES

Evolution is a process by which nature tinkers with the parts of existing organisms rather than inventing new ones. Because evolution cannot "see" into the future, changes must be of immediate use. Wholesale reinvention of functions is simply not possible. As a result, there are many examples of complex structures that performed one function being gradually modified to do something entirely different. Often it is clear that a brand-new structure might have been better, but natural selection works only on tiny modifications, like a house being endlessly remodeled to accommodate the changing number and lifestyle of its occupants rather than bulldozed and rebuilt. There are thus countless examples of ancestral forms evolving slowly over millions of years into different species, adapted to different habitats. These similarities provide powerful and easily visualized evidence for evolution.

One striking example is the way our hands and feet are so similar to the forelimbs of other mammals. The similarities with the orangutan and other primates are obvious; what is not so obvious are the similarities with the bat and the mouse. When we compare these species, we might expect to discover entirely unrelated mechanical configurations of bones. After all, what we do with our appendages bears little resemblance to what bats do with theirs. What we find, however, is the same configuration modified for different purposes. In all these cases we find creatures with five "fingers" (or toes, if you prefer), each of which is segmented into digits. In bats the digits are dramatically extended to make a frame for a large unfolding wing; in mice they are smaller and closer together, adapted to walking; in humans, they are optimized for complex motor skills. There is nothing magic about the number five and yet all these creatures have the same number of "fingers." One has only to ponder one's little toe to realize that, for many applications, it would have been fine for "one little piggy" to have gone to market and not come back. Evolution offers the compelling explanation that a five-fingered ancestral form passed down this property to a large number of species. Natural selection worked within this constraint to give us the many variations around this common theme that we see today.

The details of such processes are illustrated in the way that the mammalian jaw evolved from its reptilian ancestor. Mammalian jaws have a single bone, whereas reptilian jaws contain several. (This allows them to open their mouths so wide they can eat things almost as large as they are. It also makes them good inspiration for creepy space aliens.) In the path from reptile to mammal, well documented in the fossil record, the "extra" reptilian jaw bones gradually move back in the head and become the hammer, anvil, and stirrup found in the mammalian ear. These connections explain why stretching your jaw often "pops" your ears.

Without the explanation of common ancestry, similarities like these would be deeply mysterious. Why would the bones in our ears resemble those in the jaws of reptiles? Evolution answers this question.

DEVELOPMENTAL SIMILARITY

Two-month-old embryos of chicken, pigs, fish, and humans look similar. They all have gills, webbed hands and feet, and tails. In a few weeks these formations disappear from the human embryo. What is going on? This fascinating puzzle has a simple evolutionary explanation.

The fish is the oldest of these four species and keeps all these formations into adulthood. The human is the most recent and keeps none of them. Pigs keep their tails; chickens hang on to the webbed feet and the tail, but lose the gills. The evolutionary explanation is that the fish is the common ancestor of all three, with genetic instructions to bring these formations into full maturity. But natural selection, as it tinkered with the transitional forms between the ancient fish and the more modern mammal species descended from the fish, found it easier to shut down various formations in the womb (or egg) rather than remove the genetic instructions that give rise to them. Occasionally, however, this shutdown gets derailed and human babies are born with webbed hands and feet.

Obviously, if human babies are sometimes born with webbed feet and hands, the human genome must have instructions for this. It is quite unlikely that a genetic defect could result in the production of webbed feet from nothing in a single generation, as if the entire set of instructions to do this somehow appeared by accident out of nowhere. Far more likely is that a shutdown instruction got disabled, resulting in full production of the undesired webbing.

Embryology studies like these impressed Darwin, even though he knew nothing of the simple genetic explanation that would eventually be provided for the phenomenon: "How, then, can we explain these several facts of embryology,—namely the very general, though not universal, difference in structure between the embryo and the adult;—the various parts in the same individual embryo, which ultimately become very unlike and serve for diverse purposes, being at an early period of growth alike?"[15]

The answer Darwin provided was *common descent*, also known as common ancestry. It is an insight that continues to receive compelling confirmation as the genomes of various species are mapped and compared.

EVIDENCE FROM GENETICS

Because multiple independent lines of evidence support evolution, it is instructive to compare the conclusions on one line of evidence with another. If, for example, the evidence from paleontology doesn't line up with the evidence from genetics or biogeography, then something is wrong. But when *independent* lines of evidence converge, like in a rock-solid court case, the conclusion becomes quite irresistible.

DNA studies are the most recent line of evolutionary investigation and, as such, work within the preexisting framework provided by the other approaches that have been around longer. Not surprisingly, ongoing DNA studies are steadily clarifying and confirming the general evolutionary picture. Embryology, for example, has been powerfully augmented by DNA studies. We now know that the DNA sequences of humans and other species, even nonprimates, are very similar and share many instructions in common. As more and more genomes are mapped, an increasingly clear picture of the trajectory of life will emerge, including the specific genetic changes that gave rise to new species.

The genes we share in common with worms, for example, contain coded instructions about the most primitive structural elements of our bodies, such as basic body segmentation—getting our head at the right end—or orientation—keeping the back and front from getting mixed up. The genes we share with dogs include instructions for spinal formation. The genes we share with more recent ancestors code for more distinctive features, like our complex brains and remarkable hands.

Ongoing research on the genomes of various species is turning up far more commonalities than even the most enthusiastic evolutionist might have predicted. Not long ago it was conventional wisdom that the eye had evolved many times independently.[16] But recent studies indicate that the genetic instructions for the eye are shared by many different species, from fruit flies to humans. Current genome studies are providing dramatic evidence that we share much of our biology with other species.

Recent studies have established that, in addition to sharing genes that do useful things, like making eyes or hemoglobin, we also share nonsense genes with other species. Called pseudogenes, these bits of "misspelled" DNA make such a compelling argument for the reality of common ancestry that one leading evangelical biologist claims they establish common ancestry as a fact.[17]

A pseudogene is a piece of mutant DNA that has no obvious function, often because it sits beside a healthy unmutated version of itself that does the work it is supposed to do. Because pseudogenes don't actually do anything, there are no selection pressures to remove them—they don't interfere with reproduction—and they can pass securely from parent to offspring, from parent species to daughter species, across millions of years.

A pseudogene is like a misspelled word in a book. When I was a college student I worked as a teaching assistant, grading astronomy homework. On one homework set a student wrote about the "protons" that travel to earth from stars. I scrawled something uncharitable in the margin, explaining the difference between "photons," which do come to earth from stars, and "protons," which most certainly do not. I was quite puzzled when the next student made the same mistake, and I suspected cheating. When a third student made the same mistake, I decided that this wasn't a coincidence and, as the number continued to rise, I decided to consult the textbook. Sure enough, there was a typo in the discussion of how stars shine, referring to the production of *protons,* when the word should have been *photons.* The common mistakes in the homework all had a common ancestor in the textbook. Can we possibly conclude otherwise? Likewise, when we find the same pseudogene in many different species, how can we conclude that this identical genetic misspelling happened many times?

There are other forms of genetic gibberish that can be handed down as well and used to trace ancestry. Retroposons are strings of nonsense DNA that are readily overlooked when the genes are being read, just as the string of gibberish jkjkjkjkjkjkkjjkjkkjj can be easily overlooked in reading this sentence. If this preceding sentence were quoted in twelve otherwise different reviews of this book, we could hardly believe that the identical string of gibberish emerged independently in all twelve cases.

Identification of retroposons has secured the evolutionary inference, based on fossils, that the whale is related to the hippopotamus, cow, sheep, deer, and giraffe, all of which are "even-toed ungulates." An identical piece of genetic gibberish appears in all of them, at exactly the same place in their genomes. There is simply no explanation other than an original appearance of this genetic string in a common ancestor.[18] If the theory of evolution did not exist today, such discoveries would compel scientists to develop it to explain data like this.

The single most dramatic commonality of life on this planet, of course, would have to be the very chemistry of our genome. Without exception, the genomes of every species, from poison ivy to chimpanzees, use the same DNA language, which has just four "letters"—the molecules cytosine, guanine, adenine, and thymine. These four molecules, called nucleotides, are typically referred to by their first letters: C, G, A, and T. There are other nucleotides that would work equally well—perhaps even better—but somehow *every* life-form on the planet has its genetic code written in this particular language.

To appreciate the significance of this, imagine an alien anthropologist studying humans and discovering that, although humans speak many different languages, in every country there are large numbers of people who speak English. The existence of multiple languages establishes that human communication does not have to occur in English, so our alien anthropologist certainly can't infer that English possesses some strange feature ensuring that every time a language is developed, it will be identical to English.

Our alien anthropologist would have to infer that the English spoken in every country is derived from a single source, a linguistic common ancestor.

CONCLUSION

Evolution unites the disparate data surveyed above in ways that creationism simply cannot. If creationism were true, we should be able to explain the facts of biogeography in terms of animals and plants dispersing from Noah's ark or at least radiating out from the Middle East in some way. We cannot. If creationism were true, we should be able to find some explanation for pseudogenes other than common ancestry. If creationism were true, we should be able to explain the sequences in the fossil record without invoking billions of years of natural history. If creationism were true, we should be able to explain why bats, mice, and humans all have five "fingers" on each "hand." We cannot.

Absent evolution, thousands of patterns in nature become completely mysterious, without explanation. Creationism offers virtually no alternative explanations, and most of its "evidence" is nothing more than a catalog of small details that don't fit neatly into the standard evolutionary scenario. Rejecting evolution on the basis of these small details, however, would be like abandoning modern medicine because it can't cure every illness or declaring that meteorology is not a science because weather forecasts are sometimes unreliable.

We must also take note of another major distinction between evolution and creation. The thousands of scientists who work within the broad paradigm of evolution—the geneticists, paleontologists, biogeographers, biologists, biochemists, and so on—all agree on the broad outlines of the theory. They all agree that common ancestry is a fact. They agree that the earth is billions of years old. They agree that many species, like the dinosaurs, went extinct long before human beings appeared. They agree that natural selection is an important process.

In contrast, the much smaller community of anti-evolution creationists and intelligent design proponents has no such shared vision of its alternative "science." Leading young-earth creationists Duane Gish, Ken Ham, and the late Henry Morris agree that the earth is young, humans and dinosaurs lived together, and the fossils were laid down by Noah's flood. They agree that the scientific evidence is solidly on their side. But their junior colleagues, young-earth creationists Kurt Wise, Paul Nelson, and John Mark Reynolds, are candid in their admission that the scientific evidence is *not* solidly on their side and that young-earth creationism is compelling primarily as an implication of biblical literalism. Old-earth creationist Hugh Ross is also a biblical literalist, but he believes both that the Bible and science agree that the earth is billions of years old and that Noah's flood was a local affair. Ross rejects all aspects of evolution, including common ancestry. Intelligent design proponent Behe rejects portions of evolution, but accepts common ancestry and the great age of the earth. In his most recent book he makes a deliberate effort to distance himself from traditional creationists.[19] Johnson, the leader of the anti-evolutionary crusade, has scrupulously avoided taking a clear position on just about anything in order to avoid dissension in the ranks that might weaken the collective assault on evolution.

The differences that separate anti-evolutionists from each other guarantee that they will never actually produce a real "creation hypothesis," as the title of a popular anthology suggests.[20] These differences are so great that there is simply no common ground on which to meet and resolve differences. Wise thinks the earth is ten thousand years old; Ross thinks it is five billion. Differences of this magnitude are not likely to be "ironed out" by simply sitting down together with some charts and a stopwatch. Furthermore, when the various species of creationists write about each other, they can be quite vicious. If they were not united in opposition to evolution, they would be aggressively attacking each other.

Creationism and intelligent design have thus made little progress, despite decades of huffing and puffing and blowing on the house of evolution. They continue to offer little more than a hodgepodge of anti-evolutionary microarguments, many of which date back to the nineteenth century. Were evolutionary theory to suddenly collapse, as creationists have been confidently predicting for over a century, there would be no shared "creation hypothesis" on which to build a scientific research program.

In dramatic contrast, evolutionary scientists have so many shared commitments that finding common ground on which to resolve differences is easy. The history of evolution certainly has its share of controversies, and there have been multiple small revolutions within the field. But the general agreement on the "big picture" has made it possible to negotiate these various controversies and move forward. And this, of course, is why Darwin's theory has made so much progress in the last century and a half. Evolution as an explanation for the history and diversity of life on this planet is, quite simply, true.

In the meantime, thoughtful Christians who have taken the time to reflect on evolution have found ways to make it a part of their understanding of God's creative process. In the same way that Christians made peace with Galileo's astronomy, once they stopped trying to disprove it, many have made peace with Darwin's theory of evolution. Some have even found evolution to be a rich resource for theology, a "disguised friend of faith," in the words of one thoughtful observer.[21]

PILGRIM'S PROGRESS

Every summer for the past three decades, I have made the same modest pilgrimage. In an old handmade wooden canoe I paddle to the far end of Indian Lake, an unsung body of water just outside the middle of nowhere in rural New Brunswick. The lake is long and narrow and curves around at the end, like a finger on a baseball. Because the lake is remote and far from electricity and population centers, it is usually quiet. The few cabins clustered at one end are generally unoccupied, and my canoe typically has the lake to itself.

The trip to the end of the lake takes about an hour, depending on the wind and, most recently, how vigorously my daughter in the back of the canoe is willing to paddle. As the canoe moves around the bend in the lake, the cabins on the other end recede from view and, with their passing, all indications that Indian Lake shares its pristine wilderness with human beings disappear.

Underwater springs and a modest stream cascading down the hillside feed the lake. The height of the water, which varies, is determined by the beavers that dam up the area where the water runs off into a marshy forest. Once, when a particularly industrious family of beavers took up residence there, I had to tear out some of the dam to lower the level and keep the water from encroaching on my cabin.

A tiny population of loons observes my pilgrimages to the end of the lake. Their mournful laugh authenticates that this is true wilderness, for they are threatened by powerboats and waterfront development. Avid swimmers, the loons dive under the water when startled and reappear somewhere else on the water; I am far more interested in them than they are in me.

The goal of my pilgrimage is always the same—to sit quietly and motionless near the beaver dams and hope the beavers come out.

I love listening to the wilderness—the whisper of leaves, the cascading water, the strange harmony of the birds. It's an interesting sound, almost driven to extinction today by cell phones, televisions, iPods, and planes flying overhead. No planes fly over Indian Lake, though, because it is on the way to nowhere.

I feel strangely at home in that canoe with my daughter. I sense some approval of my presence here among the lily pads just outside the beavers' huts. I feel connected to this little bit of landscape that I have visited every summer since before my children were born. I note that the beavers build their primitive dwellings at this end of the lake, while my species build ours at the other. The beavers construct their homes as a protective haven for their offspring; I suspect they feel about their children somewhat as I do about mine, minus the fretting about boyfriends and college expenses. Little beavers count innocently on their parents just as my daughter behind me in the canoe counts on me.

The wilderness experience is therapeutic for reasons we don't understand very well. Human beings prefer a landscape of trees and lakes to skyscrapers and parking lots. Waterfront property commands a huge premium as real estate; homes are consistently situated to take advantage of natural beauty; and properties are landscaped to look as natural as possible. Everyone agrees that a meadow or a pond is more beautiful than a parking lot, and we will pay well to avoid looking at the parking lot. But why? Why do we all agree that brightly colored cars on gray asphalt are unattractive, but colorful flowers in a green meadow are beautiful?

Research shows that people who drive along tree-lined roads arrive at work with lower blood pressure than those who commute along streets lined with buildings. We nurture plants indoors to soften the artificial character of our homes and workplaces. Owning a pet increases our life expectancy.

We are connected to the natural world in so many ways and, though nature is sometimes "red in tooth and claw," often it is not. As Darwin wrote so eloquently at the end of *On the Origin of Species:* "There is grandeur in this view of life, with its several powers, having been originally breathed into a few forms or into one; and that, whilst this planet has gone cycling on according to the fixed law of gravity, from so simple a beginning endless forms most beautiful and most wonderful have been, and are being, evolved."

Darwin joined all of life together in a most magical way and in so doing dismantled the wall that separated humans from the rest of nature. Critics of Darwin warn ominously that he has reduced human beings to the level of the animals and this accounts for our supposedly bad behavior of late. But this is the "glass half empty" perspective. Might we not say instead, and more optimistically, that Darwin has raised the level of the animals? Darwin provides for us a new appreciation and respect for the loyalty of our dogs, the devoted attention of the mother bird, the industry of the beaver, the playful spirit of the otter, the proud countenance of the wolf, the human-like curiosity of the higher primates. Darwin may have closed the gap between humans and animals, but he did that by promoting the other species, not demoting ours.

We met the bonobo Kuni in the Introduction to this book. Kuni demonstrated great compassion and intelligence in caring for a troubled bird. Kuni's attentive kindness went beyond what most humans would have done. When a bird stuns itself by crashing into my window, I do little more than set it at the edge of the woods, if I do anything at all. I certainly don't hover over it to keep predators at bay while it recovers.

Kuni can't do calculus, and I can. Kuni can't play the guitar or write a book. By the yardsticks we typically use I am superior. But we are learning that intelligence should not be measured along a single yardstick. Kuni's demonstration of interspecies compassion inspires me in ways I find provocative.

THE GLASS HALF FULL

Nature is grand on so many levels. Does this grandeur have something to do with the fact that it was created by God? There is an artistic character to nature that has always struck me as redundant from a purely scientific point of view. Although I am a scientist and a great enthusiast for that approach to understanding the world, I often find myself thinking that our scientific understanding is an inadequate abstraction, that only a portion of reality has been captured in its nets. And maybe that portion is smaller than we think.

I am interested in knowing why it is so intriguing to watch the birds outside my window. Why do they sing so much? Why is their song so pleasant for humans to hear? Why, for example, does almost every scene of undeveloped nature seem so beautiful, from mountain lakes to rolling prairies? If the evolution of our species was driven entirely by survival considerations, then where did we get our rich sense of natural aesthetics? Perhaps there are answers to these questions. E. O. Wilson has coined the word *biophilia* to describe our affinity for nature and started some tentative explorations in these directions.[1] But I wonder how far those explorations will take us.

The scientific approach to nature is strongly biased in favor of engineering analogs, the legacy of Newton's mechanical view and the great power of mathematics. We tend to view the eye as an optical device, the brain as computational, and the knee as mechanical. We borrow what understanding we can from these metaphors. Phenomena without engineering analogs, like our sense of humor or great enthusiasm to play in rock bands, seem harder to understand. I worry that scientific progress has bewitched us into thinking that there is nothing more to the world than what we can understand. Science is like the fisherman's net that can't catch small fish because the holes in the net are too large. We must be careful not to conclude that the fish we can catch disprove the existence of those we cannot. Our failure to understand the deep aesthetic of nature must not delude us into thinking that it does not exist or that the meaning we derive from it is illusory.

The challenge for the religious believer is, of course, the claim that God created everything and whether the grand tapestry of nature can be described as God's handiwork. I side with Darwin in rejecting the idea that God is responsible for the details.[2] There are too many things that don't fit into the standard creationist scenario—bad design, instinctual cruelty, pointless waste. On the other hand, anyone who has contemplated nature in any detail comes away with a deep appreciation for its rich creativity. I am attracted to the idea that God's signature is not on the engineering marvels of the natural world, but rather on its marvelous creativity and aesthetic depth.

Scientists are not supposed to talk about God in this way, for it raises questions that can't be answered. And it upsets Richard Dawkins. But I am going to do it anyway. And I know that the intelligent design theorists I have dismissed earlier will accuse me, perhaps with justification, of being a hypocrite for rejecting the way they talk about God, but then offering my own version of God talk.

THROUGH A GLASS DARKLY

Darwin offered us two revolutions. The first was the destruction of the traditional creationist picture, where God created all things via individual supernatural acts more or less as we find them today. This revolution is one we must accept, despite ongoing hostility from conservative Christians. There is simply too much evidence in its favor. Darwin's second revolution was the establishment of random, "purposeless" selection processes, natural and sexual, as the only creative mechanisms at work in natural history. Gould, that most eloquent of evolutionists, put it like this: "We are glorious accidents of an unpredictable process with no drive to complexity, not the expected results of evolutionary principles that yearn to produce a creature capable of understanding the mode of its own necessary construction."[3]

I agree with Gould that we are "glorious," but I am not convinced we are "accidents." And I am certain that we should not be proclaiming confidently that natural history is a meaningless trajectory. Questions about the movements of history, like questions about ultimate origins, are highly speculative. Most who proclaim on such global questions offer nothing more than their personal ideology, for science has little to say about nature on that scale.

The confident assertions of evolutionists can give the misleading impression that we know everything we need to about the historical details of the process. This is simply not true. Evolution is what we call an *underdetermined* theory, which implies that many of the details are missing and have to be filled in by "connecting the dots." This underdetermination provides no argument that evolution is a false theory or so weakly supported that rational people should withhold support. It suggests, rather, that we should be careful about making global generalizations about evolution.

Dawkins, in *The Ancestor's Tale*, compares evolution to a pilgrimage, a suggestive geographical metaphor. Let me use that metaphor in a different way to illustrate the nature of underdetermination. Suppose you have the passport of a world traveler, filled with stamps from various countries and the dates and times of each border crossing. Suppose also that you have no reason to suspect that any of the information is false or created to mislead. Your task is to reconstruct the travel history of this person, using the passport and your general knowledge of how reasonable people travel.

You would naturally begin by lining up the dates and countries. You might discover, for example, that England, France, Belgium, Germany, Canada, Australia, and Brazil were visited in that order. This part of the history would be quite certain, a solid framework on which to hang additional details. Your knowledge of travel would allow you to infer that the trip from England to France was probably by train, based on the times on the stamps. So our traveler must have used the Chunnel. Further analysis of the time stamps might suggest that all the European travel was by train. This inference would be also be supported by your knowledge of Europe's excellent train system. The trip from Germany to Canada, however, was obviously not by train. Since it took one day, you conclude that it must have been by plane, rather than boat. For similar reasons you conclude that the trip from Canada to Australia must have been by boat, since the travel time appears to have been several weeks and there is obviously no train or highway between those countries.

At a certain level of detail this travel history could be confidently constructed and embraced with a high level of confidence, barring some unlikely scenario like our traveler being a spy. Likewise the history of life on this planet can be constructed with a high level of confidence.

Certain questions related to our world traveler, however, would be hard to answer. Based on his passport alone, we can't infer very much about where he traveled inside France. Did he travel by train or taxi? Did he visit Paris? We might know he was in France for four weeks, but that would be it, without additional information. And why is he traveling so much in the first place? The sequence of countries he visited seems almost random, but we are certainly not justified in concluding that our ignorance of his purpose constitutes evidence that he had no purpose. In fact, most reasonable people would suppose that there was an unknown purpose and start looking for it, rather than conclude there was no purpose. Our traveler's itinerary is *underdetermined*. There is a level of detail that we simply cannot access with the information at hand.

The long pilgrimage through time that Dawkins calls the ancestor's tale is similarly underdetermined. The solid evidence from fossils and genes are the stamps in our passports. We know our story began with simple one-celled life-forms. We passed from fish to amphibians to reptiles and birds. We know our most recent history was mammalian, and our last major sojourn was on the grasslands of Africa.

But we don't know all the intermediate species through which we passed in getting from fish to amphibians. Most species appear and go extinct without leaving so much as a single fossil. The fossil record is the story of unusually successful species that lived long enough to leave behind a record of their existence. They are the "border crossings" in our passport, providing definite answers to some questions while leaving others unaddressed.

To claim, with Gould and Dawkins, that our evolutionary history, our ancestor's tale, is a long and glorious accident is to make a statement about details that we don't have. We must resist the tendency to turn our ignorance into a conclusion. Absence of evidence is only rarely evidence of absence.

Gould and Dawkins are smart guys and know way more about evolution than I do. But they are also agnostics and thus have no choice but to deny any overarching purpose to natural history. In their view, there is no traveler, just a globe-trotting passport wafting on the breeze, getting hit with the occasional stamp. They may be right, of course, but let us admit that their guess is no better than mine.

We can agree, perhaps, that the inspection of natural history per se provides no certain indication that we are the "expected results" of some hidden patterns. However, there is another way to look at this. As a believer in God, I am convinced *in advance* that the world is not an accident and that, in some mysterious way, our existence is an "expected" result. Thus, I do not look at natural history as a source of data to determine whether or not the world has purpose. Rather, my approach is to anticipate that the facts of natural history will be compatible with the purpose and meaning I have encountered elsewhere. And my understanding of science does nothing to dissuade me from this conviction.

Religious believers, from spiritual agnostics like Einstein, to people who enjoy canoeing, to enthusiastic missionaries like Mother Teresa, have always found layers of meaning in the world. Einstein in his ivory tower was enthralled by the mystery of the world's rationality and its accessibility to human reason. How can it possibly be that the brain of a creature that evolved on the grasslands of Africa can penetrate the deep secrets of relativity? Mother Teresa, in the gutters of Calcutta, experienced such a powerful compulsion to help the hopeless that she was energized to dedicate her life to bringing dignity to the most destitute of our species. In my annual pilgrimage at Indian Lake I am drawn to experience the pristine wilderness and share it with my family.

The meaning we encounter in the world is deeply mysterious and existential. In a strange way it is just *there*. Meaning is not derived or inferred from our understanding of the world. Nobody inspects evolutionary theory to see whether the world has meaning. Nobody learns Darwin's theory and decides they should have children, or live on a beach, or seek random sexual encounters, or take a safari, because billions of years of natural selection have ensured that these experiences will be meaningful. Many people, in fact, decide *not* to have children and find meaning in the opportunities enabled by avoiding the obligations of parenthood. Most of us choose *not* to live on beaches. We control our sexual urges and watch nature shows when we feel like taking a safari. The meaning we find in the world is simply *encountered* in all its rich mysterious complexity and in the most surprising of places. Meaning is not the conclusion of a scientific investigation.

Our religions are responses to both our need for meaning and our discovery of that meaning. In the construction of worldviews that try to understand the meaning of our experience we draw connections. We gather with others who see those same connections. Religious communities are both the celebration of the meaning we find in community and the expression of our biological need to be in community. The praise we offer for the transcendent beauty of nature is both a response to that beauty and a celebration that we are creatures capable of that response.

The connections we draw between the mysteries of our existence and those parts of the world we understand must be drawn always in pencil, in anticipation of being erased. Once we connected our relevance to a location in the center of the universe. There were observations that supported this cosmology, to be sure, but the enduring power of this view derived from its connection to an exalted view of ourselves. That line had to be erased. For much the same reason we once celebrated a large gap between our species and all others. That line has also been erased.

Science has perhaps gotten as much from the materialistic paradigm as it is going to get. Matter in motion, so elegantly described by Newton and those who followed him, may not be the best way to understand the world. Science has moved into an age of information, a new and productive way of looking at the world and encouraging in many ways. Human beings are the most complex creatures in the known universe, and perhaps, once again, we find ourselves on the top or at the center or on the motherboard or wherever we locate the privileged spot in an age of information.

Charles Darwin challenged the traditional view of creation, to be sure. But the facts of nature were challenging the traditional view long before Darwin came along, if only we had been willing to look more closely. Nature has always been an untidy, bloody affair. Its messiness is not easily reconciled with the traditional idea of a creator, unless one could be satisfied with the odd explanation that cats torture their prey and roses have thorns because Adam and Eve sinned. Many thinkers welcomed Darwin, seeing in evolution a more acceptable role for God as creator.[4] And there continue to be thoughtful Christians today who have made their peace with evolution and are doing creative theology as they explore what it means to be biologically evolved creatures in relationship with God. That so many other Christians are now massed against Darwin has more to do with American culture than biology or Christian theology. From a purely theological perspective there is much in evolution to interest Christians, and if controversy had not driven discourse so quickly off the rails, the voices of reason might today be speaking from the center of American Christianity, rather than the fringes.

THE QUESTION

Today as I was leaving class a thoughtful student approached me and wanted to know if I was going to "come clean" about evolution and let the students know what I believed. I had been lecturing on Darwin, trying to get the students inside the great scientist's head as he wrestled with the observations that eventually led him to the theory of evolution. This student, like me, was raised to believe that Darwin was evil and evolution was a lie. But, also like me at his age, he was having second thoughts as he was becoming better informed (or brainwashed by his professor, depending on your perspective).

When I teach Darwin, I avoid taking a position, partly so students can feel free to reject evolution if that is their choice. More important, though, I want the students to wrestle, as Darwin did and I did when I was their age, with the implications of cruelty in nature and bad design. They need to confront, on their terms, the mass of data that can't be reconciled with the Genesis creation accounts. If I lay my position out too clearly, some students will make their decision based on what they think of me, rather than the issues at stake.

Many college students, and most Americans for that matter, have little interest in evolution as science. Their concern is that science not crowd out their religious beliefs. At some level they fear Daniel Dennett's "universal acid" may actually have the power to dissolve their beliefs. And they don't want to find out if that is true.

Their fear is understandable. Almost everyone who talks about evolution insists that we must make a choice between evolution or creation, materialism or God, naturalism or supernaturalism. Dawkins and Dennett believe this and say, "Choose evolution"; Johnson and Morris believe this and say, "Choose creation." The four of them are grand evangelists for the positions they have chosen. Just as significantly, all four of them are champions of a false dichotomy.

This dichotomy plays well in the press. It's controversial, combative, and simple. There are good guys and bad guys, no matter where you stand. But this dichotomy is *wrong*. These are not the only two options. These are not even the most reasonable options.

COMING CLEAN

I think evolution is true. The process, as I reflect on it, is an expression of God's creativity, although in a way that is not captured by the scientific view of the world. As soon as we start highlighting specific places where we think we glimpse God's handiwork, we open ourselves up to the old "God of the gaps" problem. I think there are ways, though, that we can begin to look at the creation and understand that the scientific view is not all-encompassing. Science provides a partial set of insights that, though powerful, don't answer all the questions.

Intelligent design and scientific creationism seem inadequate to me, because they reduce God to one agent among other agents in natural history. If ID is true, then it implies that the agents of evolution are natural selection, sexual selection, God, mutation, chance, and whatever else you want on the list. Each of these agents makes its own individual contribution. Natural selection made saliva, God made hemoglobin, sexual selection made the peacock's tail, and chance drove the dinosaurs to extinction. God is one of several agents of change. God may be the "big gun" who steps in to do the projects that exceed the capacities of the other agents, perhaps, but God is still just one agent among many. Is this really how we want to think about natural history?

God's creative activity must not be confined to a six-day period "in the beginning" or the occasional intervention along the evolutionary path. God's role in creation must be more universal—so universal that it cannot be circumscribed by the contours of individual phenomena or events. We must resist the temptation to make God into a "superengineer" or "master crafts-man" or "grand artist." God may indeed have all of these attributes, but we ought not to suppose that any of them capture more than the tiniest intuition about God's role in creation. It seems to me a more hopeful perspective to step back as far as we can and examine the biggest possible picture in the hopes of getting a glimpse of what it means to say that God created the world.

A BRIEF HISTORY OF EVERYTHING

Natural history is richly layered in surprising ways. At the deepest level of reality the world is so simple it boggles the mind. There are only four kinds of interactions that occur in nature: gravitational, electromagnetic, strong nuclear, and weak nuclear. Every event, from a thought in your head, to the chirp of a bird, to the explosion of a distant star, results from these four interactions.

There are only two kinds of physical objects in the world: quarks and leptons. The familiar protons and neutrons are composed of quarks; the electron is the best-known example of a lepton. Every physical object, from a guitar string, to the Mona Lisa, to Pluto (whatever it is these days), is made from quarks and leptons.

All natural phenomena, no matter how rich or mundane, result from two kinds of particles interacting via four kinds of interactions. Who could possibly conceptualize the extraordinary creativity of a world built like this?

Imagine that we have a film of the entire history of the universe. Titled "Four Forces and Two Particles," the film offers so little promise that you can hardly bring yourself to pay attention. "Boring," you note to yourself, expecting to see nothing but zillions of marbles floating in the blackness of space.

The world you think will be so boring starts to unfold, and you sit back to watch. In no time at all things start to happen. The quarks, which have electrical charges of 2/3 and −1/3, start sticking together under the influence of the strong nuclear force, and soon they are all gathered into protons and neutrons, which have electrical charges of 1 and 0. All the fractionally charged particles in the universe are gone. "Interesting," you think, "but it's still just marbles."

Protons, neutrons, and electrons are now buzzing about in a chaotic but steadily cooling mix as the universe expands. Their electrical interactions are pulling the electrons toward the protons. As the temperature declines across a certain key threshold, the electrons suddenly drop into orbits around the protons and the universe is full of hydrogen atoms. All the particles in the universe are now electrically neutral and you note something rather puzzling you had missed: the universe has a perfect balance between the positive and negative charges. Interesting ...

Now that all the particles are electrically neutral, the powerful electrical force stops dominating and the much weaker gravitational force takes over. The hydrogen atoms are pulled ever so slowly together until they begin to cluster. Steadily growing balls of hydrogen form and, as the balls get bigger, their gravitational pull on other atoms increases until much of the hydrogen collects into huge balls. The balls get steadily larger, surpassing the size of the moon, then the earth, then a large planet like Jupiter. This, of course, is more interesting, but it is still just marbles.

Suddenly, another critical threshold is crossed, and the balls ignite. Like a drawn-out fireworks display, all across the universe great balls of hydrogen turn into stars. Things are certainly more interesting than you had anticipated, but the universe is still boring, composed almost entirely of hydrogen.

The gravity within these newly born stars crushes the hydrogen atoms like eggs under a steamroller. Astonishingly you notice that the strong force cooperates intimately with this gravitational crushing, and the hydrogen atoms combine to become helium atoms. In fact you notice that the very process that emits the light from the stars builds the periodic table. Hydrogens make helium. A helium and a hydrogen make a lithium. Two helium make beryllium. A beryllium and a helium make a carbon. Other combinations make nitrogen, oxygen, neon, sodium, and on down the periodic table. Interesting ...

A large star suddenly explodes with the force of a billion atomic bombs and brings you out of your seat. The explosion fills a massive region of space with the elements created inside the star; the powerful explosion, though, is strangely orderly. Gravity starts gathering the debris back into balls again, and a large chunk at the center becomes another star. This time, however, some of the balls end up orbiting about the second-generation star. These smaller balls have a rich roster of elements, since they formed from the debris of a star that had converted much of its hydrogen into other elements. In particular, many of these smaller balls possess a curious molecular combination of hydrogen and oxygen. In most parts of the universe, these molecules are in the form of a solid. In the others they are a gas. But on balls that are exactly the right distance from the central star, the molecules are liquid, a particular liquid called *water*. Very interesting ...

The universe is now anything but boring. You marvel at the complex structures that have been built from the simplest of raw materials. Water turns out to be especially amazing and surprisingly capable of encouraging the formation of ever more complex molecules like amino acids, proteins, and enzymes. These complex materials build up steadily until, to your astonishment, a particular arrangement actually starts duplicating itself and the waters become filled with this new process. In amazement you realize that the universe now has life. This is not boring at all.

Subtle and highly nuanced interactions between these primitive life-forms, driven by a seemingly endless set of molecular interactions, steadily and mysteriously push the life-forms to greater and greater complexity. Some kind of subtle advantage accompanies increases in complexity. It is almost like a mysterious force calling life—and the universe—to become ever more interesting. And yet you know that all this is simply the way that quarks and leptons combine under the influence of the four interactions. Your earlier fears that the universe was going to be boring and meaningless now seem laughably ill-founded.

The film turns out, upon reflection, to be quite extraordinary. The credits begin to roll, blurred and illegible, as you watch a member of your own species paddle a canoe to the end of a small lake with his daughter and then sit quietly in a pristine wilderness. Somehow a universe that started out so hopelessly boring has turned out to be quite interesting after all.

MORE THAN PARTICLES

The trajectory of natural history leading to human beings is an amazing story. As we look back at the earlier stages of the universe, we cannot help but marvel at how later developments build on previous ones. There is an "unfolding" to the process, as if each stage is both the completion of what has gone before and the anticipation of what is soon to come. Freeman Dyson, one the greatest scientists of the twentieth century, puts it like this: "The more I examine the universe and study the details of its architecture, the more evidence I find that the universe in some sense must have known that we were coming. There are some striking examples in the laws of nuclear physics of numerical accidents that seem to conspire to make the universe habitable."[5]

The tools of science have been effective at illuminating each individual step on this long and winding road. But many things exist in nature that science does not even try to explain. Those are labeled "chance." When a scientist claims that something occurred by "chance," that is an admission that there is no explanation. I hasten to point out that this does not mean that some causal factor is missing and has to be provided by God. What it does mean, though, is that events occur in nature that fall outside the explanatory purview of science. These events are either genuinely without explanation or to be explained from a perspective outside of science. This offers no proof of God's intervention, of course, for it may indeed be that the events are without explanation. But such inexplicable aspects of creation at least erect explanatory boundaries for science and preclude global generalizations about what it all means.

I mentioned above that the story of evolution was underdetermined by virtue of the large portion of the story that is simply missing. In a more profound way, all of nature is underdetermined. The natural order, as disclosed so remarkably by contemporary physics, is not a closed system of interlocking mechanical parts, as the Newtonian worldview mistakenly implied. Rather, events unfold in ways that are not entirely specified by the laws of physics. The most famous statement of this underdetermination is the Heisenberg Uncertainty Principle, a precise mathematical articulation of exactly how much we can and cannot know about the world. In one of the most general laws in all of science, Heisenberg's principle sets clear limits on how accurately we can know the behavior of particles, such as electrons. An electron passing through a small hole, for example, will have its path altered by the interaction with the hole. We can know that the trajectory will bend by a certain amount, say ten degrees. But we cannot know in which direction; whether the electron goes left or right, up or down is determined by "chance."

If reality contains nothing but quarks, leptons, and four interactions, then history is indeed filled with chance events and, as Dawkins and Dennett would have us believe, we are the result of a mindless process. If God exists, however, then other possibilities open up. Perhaps the unfolding of history includes a steady infusion of divine creativity under the scientific radar. Perhaps the meaning we encounter in so many different places and so many different ways is not simply an accident of our biology, but a hint that the universe is more than particles and their interactions.

Anthropologists tell us that our forebears bowed before the mystery of their poorly understood and intimidating world. Mother Nature was capricious and unpredictable, to be worshiped, patronized, and feared. Animals and even children were sacrificed to the gods to hold their wrath in check. We have now explored this world and discovered that we need not fear Mother Nature, at least not in the same way.

In coming to understand the world so much better, however, we have not banished its mystery. We may not cower before the lightning bolt or hide from the thunder. We no longer throw virgins into volcanoes to appease the lava god. Now we bow before different mysteries. We float tenuously on a tiny planet in the immensity of space, humbled by the knowledge that it took ten billion years of cosmic evolution to prepare our planet for life. Four billion years of the most remarkable and creative explosion of life has preceded us on that planet. And now, here we are, at once both a fragile species and, strangely, a danger to other species and even to ourselves. We have learned so much about the inner workings of our world, and yet so little about what we should do with that knowledge. In deep and important ways we have not dispelled the mystery of our existence at all—we have simply established it with greater clarity.

"We shall not cease from exploration," wrote T. S. Eliot in "Little Gidding," "and the end of all our exploring will be to arrive where we started and to know the place for the first time."

ACKNOWLEDGMENTS

This project began with a phone call out of the blue from Ken Giniger, asking if I knew anyone who might be "interested in writing a book on the evolution controversy for HarperOne." Having just completed a book on a related topic and wondering about my next project, I hastened to recommend myself for this task. Ken fast-tracked the approvals and now, less than two years later, the book is finished. I must thank Ken for conceptualizing this project, giving it to me, and then offering many valuable suggestions on the final version.

My editors at HarperOne, Kris Ashley and Mickey Maudlin, provided valuable input that helped me get started on the right foot. And Kris read the entire draft when it was completed and made many useful suggestions for improvements, all of which I sensibly heeded. Despite my careful scrutiny of the manuscript, my copy editor at HarperOne, Ann Moru, identified countless literary faux pas and made many valuable suggestions for improvement.

I owe a special thanks to several of my busy colleagues at Eastern Nazarene College. Eric Severson took the time to read the first draft of the entire manuscript for both style and content. A consummate teacher, Eric couldn't help making countless suggestions, and the finished product is significantly stronger on several levels because of his efforts. Donald Yerxa and Randall Stephens read a later draft and made valuable suggestions. And Karen Henck and Kelsey Towle, who teach writing at my college, read portions of the manuscript and offered helpful advice on style.

Ron Numbers and Jon Roberts, both leading—and very busy—scholars for whom I have enormous respect, graciously agreed to read the entire manuscript for accuracy. Not surprisingly, they uncovered a number of problems that I hope I have addressed adequately. Both Numbers and Roberts have done important work in this field, and it was gratifying—and somewhat intimidating, to be honest—to have them review my manuscript. Watching the Red Sox at Fenway with them, however, was just plain fun.

Francis Collins, who has to be one of the busiest scientists on the planet, read *Saving Darwin* on a plane ride to Singapore and wrote the Foreword to the book. I am humbled by his encouraging words of support and deeply appreciative of his contribution.

I love to surround myself with talented undergraduates, and this project was no exception. For little more than minimum wage they tracked down sources, formatted footnotes, harassed my faculty colleagues with questions, copyedited, fact-checked, and generally reminded me of why it's great to be a college professor. I would like to thank them and wish them well as they begin their careers: Kurtis Biggs, Amanda Egolf, Sara Kern, and Cameron Young.

During most of the writing of this book, I was just down the hall from two exceptional writers, Marc Kaufman and Heather Wax. I would like to thank both of them for many interesting conversations about words and ideas and for helping me become a better writer.

I especially thank the John Templeton Foundation for providing financial support for this project. Its generous support released me from a heavy teaching load at the college where I work and provided resources to hire student assistants and purchase a small library of relevant books from Amazon.com.

Finally, I would like to thank my wife of many years, Myrna, for her patience and willingness to let me live in the corner of our sunroom, next to the dogwoods and the bird feeder, where I did most of the writing.

It would be nice if all this help could guarantee a flawless final product, but alas, an army of advisers can do only so much. The mistakes that remain, as the refrain goes, are mine, and mine alone.

NOTES

INTRODUCTION: THE DISSOLUTION OF A FUNDAMENTALIST

1. John C. Whitcomb and Henry M. Morris, *The Genesis Flood: The Biblical Record and Its Scientific Implications* (Phillipsburg, NJ: Presbyterian and Reformed Publishing, 1961); Henry M. Morris, *Many Infallible Proofs* (Green Forest, AR: Master Books, 1974).

2. All biblical references in this chapter are taken from the King James Version, which has long been the preferred version of the Bible for fundamentalists.

3. Romans 8:22. The King James says, "For we know that the whole creation groaneth and travaileth in pain ..."

4. Poll: "Creation Trumps Evolution," *CBS News,* November 22, 2004, http://www.cbsnews.com/stories/2004/11/22/opinion/polls/main657083.shtml. Because the poll was imprecise, we cannot distinguish between the three strands of literal creationism: the young-earth creationism of Whitcomb and Morris, day-age creationism, and gap creationism, all of which will be discussed in detail in later chapters.

5. Although some claim that the Bible does not, in a technical sense, give a date for the creation, an approximate date can be determined by adding up the "begats." If Adam was created in the same week as "the heavens and the earth," we can then extrapolate from the chronology provided for Adam's children, his children's children, and so on. There is some vagueness in this process, but most who have done this—it takes just a few minutes—come up with a date for the creation of the universe and the earth that is between six and ten thousand years ago.

6. Daniel Dennett, *Darwin's Dangerous Idea: Evolution and the Meaning of Life* (New York: Simon & Schuster, 1995), p. 63.

7. Tertullian, *De Carne Christi* (London: S.P.C.K., 1956), p. 5 (4).

8. G. K. Chesterton, *Orthodoxy* (Ft. Collins, CO: Ignatius, 1995), p. 11.

9. Frans de Waal, *Our Inner Ape* (New York: Riverhead, 2005), p. 2.

10. Howard Van Till, *The Fourth Day* (Grand Rapids, MI: Eerdmans, 1986); Darrel Falk, *Coming to Peace with Science* (Downers Grove, IL: InterVarsity, 2004).

11. Francis Collins, *The Language of God: A Scientist Presents Evidence for Belief* (New York: Free Press, 2006).

12. Ken Miller, *Finding Darwin's God* (London: Harper Perennial, 1999).

13. John Haught, *God After Darwin* (Boulder, CO: Westview, 2000).

14. Alister McGrath, *Scientific Theology: Nature, Reality, and Theory* (Grand Rapids, MI: Eerdmans, 2002–3).

15. Keith Ward, *God, Chance and Necessity* (Oxford: Oneworld, 1996).

16. Michael Ruse, *Can a Darwinian Be a Christian?* (New York: Cambridge University Press, 2001), p. 219.

17. Richard Dawkins, "Science, Religion, and Evolution," *New York Times,* May 21, 1989, sec. 7, late city final edition.

CHAPTER 1: **THE LIE AMONG US**

1. James Moore, *The Darwin Legend* (Grand Rapids, MI: Baker, 1994), p. 93.

2. Moore, *Darwin Legend,* p. 93.

3. See, for example, http://www.forerunner.com/forerunner/X0724_Darwins_Final_Recant.html (accessed March 17, 2007).

4. John Ankerberg and John Weldon, *Darwin's Leap of Faith: Exposing the False Religion of Evolution* (Eugene, OR: Harvest House, 1998), p. 127.

5. Francis Darwin, ed., *The Life and Letters of Charles Darwin,* vol. 2 (London: John Murray, 1888), p. 308.

6. Henry Morris, *The Long War Against God: The History and Impact of the Creation/Evolution Controversy* (Green Forest, AR: Master Books, 2000), p. 260.

7. Ken Ham, *The Lie: Evolution* (El Cajon, CA: Creation-Life Publishers, 1987).

8. Phillip E. Johnson, *Reason in the Balance: The Case Against Naturalism in Science, Law and Education* (Downers Grove, IL: InterVarsity, 1995), p. 59.

9. *Answers,* based in Hebron, KY, and edited by Dale T. Mason, is published quarterly by Answers in Genesis.

10. Nora Barlow, ed., *The Autobiography of Charles Darwin: 1809–1882* (New York: Norton, 1958), p. 56.

11. Barlow, ed., *Autobiography of Charles Darwin,* p. 57.

12. William Paley, *Natural Theology* (New York: American Tract Society, 1802); *The Principles of Moral and Political Philosophy* (Dublin: Exshaw, White, H. Whitestone, Byrne, Cash, Marchbank, and McKenzie, 1785); *Evidences of Christianity* (London: Clowes, 1851).

13. Barlow, ed., *Autobiography of Charles Darwin,* p. 59.

14. Barlow, ed., *Autobiography of Charles Darwin,* p. 85.

15. The importance and role of paradigms in science is convincingly articulated by philosopher Thomas Kuhn in his now classic work *The Structure of Scientific Revolutions,* first published in 1962 (3d ed., Chicago: University of Chicago Press, 1996).

16. Francis Reddy, "High-speed Star Flees Tycho's Blast," *Astronomy Magazine,* November 3, 2004, http://astronomy.com/asy/default.aspx?c=a&id=2571 (accessed March 31, 2007).

17. "The Fate of the Sun," *Time,* March 23, 1987.

18. The widespread view that science and religion have always been at odds is a fiction created in the nineteenth century by polemicists like Andrew Dickson White and John William Draper. See John Hedley Brooke, *Science and Religion: Some Historical Perspectives* (Cambridge: Cambridge University Press, 1991).

19. Isaac Newton, *Principia Mathematica* (Los Angeles: University of California Press, 1996), p. 940.

20. Thomas Paine, *The Age of Reason* (New York: Knickerbocker, 1904), p. 34.

21. David Hume, *Dialogues Concerning Natured Religion* (Indianapolis, IN: Bobbs-Merrill, 1947), p. 194.

22. Voltaire, *Candide* (New York: Bantam Dell, 1984), p. 18.

23. William Paley, *Natural Theology* (Cary, NC: Oxford University Press USA, 2006), pp. 7–16.

24. J. P. Moreland, ed., *The Creation Hypothesis: Scientific Evidence for an Intelligent Designer* (Downers Grove, IL: InterVarsity, 1994), pp. 290–91.

25. Paley, *Natural Theology*, p. 225.

26. Charles Lyell, *Principles of Geology*, 3 vols. (London: John Murray, 1830).

27. Martin J. S. Rudwick, "The Shape and Meaning of Earth History," in David C. Lindberg and Ronald L. Numbers, eds., *God and Nature: Historical Essays on the Encounter Between Christianity and Science* (Berkeley: University of California Press, 1986), p. 313.

28. Charles Darwin, *On the Origin of Species by Means of Natural Selection* (New York: Signet Classics, 2003), p. 449.

29. Quoted in Michael Ruse, *Darwin and Design* (Cambridge, MA: Harvard University Press, 2003), p. 127.

30. Quoted in Ruse, *Darwin and Design*, p. 127.

31. As mentioned, this argument is presented with surprising eloquence in Henry Morris's *The Long War Against God*. See especially chap. 4, "The Dark Nursery of Darwinism," in which Morris suggests that Darwin's fellow evolutionist Alfred Wallace was under the influence of evil spirits.

32. Charles Darwin and Francis Darwin, *The Autobiography of Charles Darwin and Selected Letters* (New York: Dover, 1958), p. 87.

33. Randall Keynes, *Darwin, His Daughter and Human Evolution* (New York: Riverhead, 2002).

34. Francis Darwin, ed., *The Life and Letters of Charles Darwin*, 2 vols. (London: John Murray, 1888), p. 56.

35. Darwin, *On the Origin of Species*, p. 459. Scholars debate the degree to which this passage reflects Darwin's true sentiments. Some would argue that this passage is best understood as Darwin playing politics and pandering to his religious critics. It is likely that Darwin's endorsements of these sentiments waxed and waned over the course of his long and tormented struggle with his faith.

36. Adrian Desmond and James Moore, *Darwin: The Life of a Tormented Evolutionist* (New York: Norton, 1991), p. 677.

CHAPTER 2: **A TALE OF TWO BOOKS**

1. Richard Dawkins, *The Blind Watchmaker* (New York: Norton, 1996), p. 6.

2. Phillip Johnson, *The Wedge of Truth: Splitting the Foundations of Naturalism* (Downers Grove, IL: InterVarsity, 2000), p. 41.

3. Richard Dawkins, *The Ancestor's Tale: A Pilgrimage to the Dawn of Evolution* (Boston: Houghton Mifflin, 2004), pp. 613–14.

4. Alister McGrath, *Dawkins' God: Genes, Memes, and Meaning of Life* (Oxford: Blackwell, 2005); Alister McGrath and Joanna Collecutt McGrath, *The Dawkins Delusion: Atheist Fundamentalism and the Denial of the Divine* (Downers Grove, IL: InterVarsity, 2007).

5. Karl Giberson and Mariano Artigas, *The Oracles of Science: Celebrity Scientists Versus God and Religion* (Oxford: Oxford University Press, 2007), pp. 19–52.

6. Johnson, *Wedge of Truth,* p. 107.

7. See http://www.the-brights.net for a description of this new movement, started in 2003.

8. David Friedrich Strauss, *The Life of Jesus Critically Examined* (London: Chapman, 1846).

9. Matthew Arnold, *New Poems* (London: Macmillan, 1867).

10. James Gibson, ed., *Thomas Hardy: The Complete Poems* (New York: Palgrave Macmillan, 2002), p. 327.

11. A. N. Wilson offers a compelling account of the Victorian loss of faith in his intellectual history *God's Funeral* (New York: Norton, 1999).

12. Christiane L. Joost-Gaugier, *Measuring Heaven: Pythagoras and His Influence on Thought and Art in Antiquity and the Middle Ages* (Ithaca, NY: Cornell University Press. 2006), pp. 48–50.

13. Louis Agassiz, "Professor Agassiz on the Origin of Species," *American Journal of Science and Arts* (June 1860): 143–47.

14. Ronald L. Numbers, *Darwinism Comes to America* (Cambridge, MA: Harvard University Press, 1998), p. 41.

15. Historian Jon Roberts provides a thoughtful analysis in *Darwinism and the Divine in America: Protestant Intellectuals and Organic Evolution, 1859–1900* (Notre Dame, IN: University of Notre Dame Press, 2001). Roberts divides the indicated time interval into two distinct phases: 1859–1875, when religious leaders inclined to reject evolution did so on the basis of its being bad science; and 1875–1900, when religious leaders, taking note of the scientific community's widespread acceptance of evolution, reformulated their objections to evolution on theological grounds, arguing that it must be rejected as incompatible with Christianity.

16. Charles Lyell, *Principles of Geology,* 3 vols. (London: John Murray, 1830).

17. John Hedley Brooke, *Science and Religion: Some Historical Perspectives* (Cambridge: Cambridge University Press, 1991), p. 237.

18. Davis A. Young, *The Biblical Flood: A Case Study of the Church's Response to Extrabiblical Evidence* (Grand Rapids, MI: Eerdmans, 1995), p. 147.

19. *The Scofield Study Bible NKJV* (Oxford: Oxford University Press, 2002), notes on Genesis 1. The Bible passages referenced are Jeremiah 4:23–27; Isaiah 24:1; 45:18.

20. John Hammond Taylor, ed., *St. Augustine: The Literal Meaning of Genesis,* vol. 1 (Mahway, NJ: Paulist, 1982).

21. Michael J. Behe, *Darwin's Black Box* (New York: Touchstone, 1996), p. 228.

22. William A. Dembski, *No Free Lunch: Why Specified Complexity Cannot Be Purchased Without Intelligence* (Lanham, MD: Rowman & Littlefield, 2002), p. 326.

23. Peter Bowler, *The Eclipse of Darwinism* (Baltimore: Johns Hopkins University Press, 1992), p. 180.

24. Ellen G. White, *The Spirit of Prophecy,* 4 vols. (Battle Creek, MI: Seventh-day Adventist Publishing, 1870, 1877, 1878, 1884, 1969), 1:85.

25. Lyman Abbott, *The Evolution of Christianity* (New York: Outlook, 1892).

26. Contributors included C. I. Scofield of the famous *Scofield Reference Bible,* Benjamin B. Warfield of Princeton Theological Seminary, James Orr of the United Free Church Collège in Scotland, George Frederick Wright of Oberlin College, and academics from many other leading Christian colleges and institutes.

27. George Frederick Wright, "The Passing of Evolution," in R. A. Torrey and A. C. Dixon, eds., *The Fundamentals: A Testimony to the Truth,* 4 vols. (Los Angeles: Bible Institute of Los Angeles, 1917; repr., Grand Rapids, MI: Baker, 2003), 4: 87.

CHAPTER 3: **DARWIN'S DARK COMPANIONS**

1. Andrew Carnegie, "Wealth," *North American Review* (June): 653, quoted in Eugenie C. Scott, *Evolution vs. Creationism: An Introduction* (Berkeley: University of California Press, 2004), p. 93.

2. Mike Hawkins, *Social Darwinism in European and American Thought: 1860–1945* (Cambridge: Cambridge University Press, 1997), p. 82.

3. Richard Weikart, *From Darwin to Hitler* (New York: Palgrave Macmillan, 2004).

4. This sordid account appears in 1 Samuel; see especially 1:2–3.

5. T. R. Malthus, *An Essay on the Principle of Population* (1798; Oxford: Oxford University Press, 2004), p. 61.

6. Herbert Spencer, *The Man Versus the State, with Six Essays on Government, Society and Freedom* (Indianapolis, IN: Liberty Classics, 1981), p. 32. The essay from which this is taken, "The Coming Slavery," first appeared in 1884.

7. Spencer, *Man Versus the State,* p. 33.

8. Francis Galton, "Hereditary Talent and Character," *Macmillan's Magazine* 12 (1865): 157–66, 318–27, http://psychclassics.yorku.ca/Galton/talent.htm. Also quoted, without reference, in Martin Brookes, *Extreme Measures: The Dark Visions and Bright Ideas of Francis Galton* (New York: Bloomsbury, 2004), p. 144.

9. Francis Galton, *Hereditary Genius* (Amherst, NY: Prometheus, 2006), pp. 305–15.

10. Edward J. Larson, *Evolution: The Remarkable History of a Scientific Theory* (New York: Modern Library, 2004), p. 193.

11. Larson, *Evolution,* p. 194.

12. George Hunter, *A Civic Biology* (New York: American Book, 1914), p. 263.

13. Hunter, *Civic Biology,* p. 263.

14. Larson, *Evolution,* p. 195.

15. Ernst Haeckel, *Natürliche Schöpfungsgeschichte* (Berlin, 1868), quoted in Weikart, *From Darwin to Hitler,* p. 106.

16. Joseph Le Conte, *The Race Problem in the South,* Evolution Series no. 29: *Man and the State* (New York: Appleton, 1892), pp. 360–61, quoted in Hawkins, *Social Darwinism,* p. 201.

17. Martin Luther, "On the Jews and Their Lies," trans. Martin H. Bertram, in *Luther's Works* (Philadelphia: Fortress, 1971).

18. Ann Coulter, *Godless* (New York: Crown Forum, 2006), p. 271.

19. Alfred Kirchhoff, *Darwinismus* (Frankfurt, 1910), pp. 73, 86–87, quoted in Weikart, *From Darwin to Hitler,* p. 184.

20. Adolf Hitler, *Mein Kampf* (Boston: Houghton Mifflin, 1943), p. 289.

21. Nazism scholar Richard J. Evans makes this point in an essay titled "In Search of German Social Darwinism": "The language of social Darwinism helped to remove all restraint from those who directed the terroristic and exterminatory policies of the regime, and it legitimized these policies in the minds of those who practiced them by persuading them that what they were doing was justified by history, science, and nature." Richard J. Evans, "In Search of German Social Darwinism: The History and Historiography of a Concept," in Manfred Berg and Geoffrey Cocks, eds., *Medicine and Modernity* (Washington, DC: German Historical Society; Cambridge: Cambridge University Press, 1997), p. 79, quoted in Weikart, *From Darwin to Hitler,* p. 233.

22. Joel N. Shurkin, *Broken Genius: The Rise and Fall of William Shockley, Creator of the Electronic Age* (New York: Macmillan, 2006), p. 194.

23. Richard J. Herrnstein and Charles Murray, *The Bell Curve: Intelligence and Class Structure in American Life* (New York: Free Press, 1994), p. 364.

24. Steven Pinker, "Why They Kill Their Newborns," *New York Times,* November 2, 1997, http://www.rightgrrl.com/carolyn/pinker.html (accessed March 4, 2007). Pinker's article has been widely criticized, but is actually much more thoughtful and exploratory than his critics would have you believe.

25. Randy Thornhill and Craig T. Palmer, *A Natural History of Rape: Biological Bases of Sexual Coercion* (Cambridge, MA: MIT Press, 2000), p. 190.

26. Nadia Mustafa, "What About Gross National Happiness?" *Time,* Monday, January 10, 2005, http://www.time.com/time/health/article/0,8599,1016266,00.html?promoid=rss_top (accessed May 17, 2007).

27. Carl Zimmer, *Evolution: The Triumph of an Idea* (New York: HarperCollins, 2001), p. 317.

28. Zimmer, *Evolution,* p. 318.

29. Niles Eldredge, *Darwin: Discovering the Tree of Life* (New York: Norton, 2005).

30. Eugenie C. Scott, *Evolution vs. Creationism: An Introduction* (Berkeley: University of California Press, 2004).

31. Ernst Mayr, *What Evolution Is* (New York: Basic Books, 2001), p. 285.

32. Ken Ham, *The Lie: Evolution* (El Cajon, CA: Creation-Life Publishers, 1987), pp. 83–95.

33. "Darwin's Deadly Legacy," television special by Coral Ridge Ministries and James Kennedy, August 26–27, 2006, on the *Coral Ridge Hour.*

34. Tom DeRosa, *Evolution's Fatal Fruit: How Darwin's Tree of Life Brought Death to Millions* (Fort Lauderdale, FL: Coral Ridge Ministries, 2006), p. 7

CHAPTER 4: THE NEVER ENDING CLOSING ARGUMENT

1. *Inherit the Wind,* videocassette, directed by Stanley Kramer, performances by Spencer Tracey, Fredric March, and Gene Kelly (MGM/UA, 1960).

2. H. L. Mencken, *Prejudices: Fifth Series* (New York: Knopf, 1926), p. 74.

3. Fredrick Allen, *Only Yesterday* (New York: Perennial Classics, 2000), pp. 166–78.

4. H. L. Mencken, "'The Monkey Trial': A Reporter's Account," *Famous Trials in American History,* University of Missouri–Kansas City School of Law, http://www.law.umkc.edu/faculty/projects/ftrials/scopes/menk.htm (accessed January 29, 2007).

5. L. Sprague de Camp, *The Great Monkey Trial* (New York: Doubleday, 1968), p. 432.

6. Clarence Darrow, *Attorney for the Damned,* ed. Arthur Weinberg (Chicago: University of Chicago Press, 1989), p. 199.

7. "Big Crowd Watches Trial Under Trees," *New York Times,* July 21, 1925, sec. I.

8. Tracy Sterling, "Darrow Quizzes Bryan; Agnosticism in Clash with Fundamentalism," *Commercial Appeal* (Memphis), July 21, 1925, sec. I, quoted in Edward Larson, *Summer for the Gods* (New York: Basic Books, 1997), pp. 190–91.

9. Edward Larson, *Trial and Error* (New York: Oxford University Press, 2003), p. 87. This chapter draws heavily on the definitive work of Edward Larson, particularly his Pulitzer Prize–winning account of the Scopes trial, *Summer for the Gods* (New York: Basic Books, 1997), and *Trial and Error,* which is the standard history of America's creation-evolution court cases up to the mid-1990s. Larson has several excellent books on the creation-evolution controversy, all of which I can heartily recommend.

10. Judith Grabner and Peter Miller, "Effects of the Scopes Trial: Was it a Victory for Evolutionists?" *Science* 185, no. 4154 (1974): 832–37.

11. Sol Tax and Charles Callender, eds., *Evolution After Darwin: The University of Chicago Centennial,* vol. 3 (Chicago: University of Chicago, 1960), p. 107.

12. *School District of Abington Township, Pennsylvania v. Schempp,* 374 U.S. 225 (1963).

13. S. McBee and J. Neary, "Evolution Revolution in Arkansas," *Life* 65, no. 21 (1968): 89.

14. "Separate Answer to the Intervention of Hubert H. Blanchard, Jr.," in Susan Epperson and Bruce Bennet, "Proceedings," in *Appendix, Epperson v. Arkansas,* 393 U.S. 97 (1968), 40, quoted in Larson, *Trial and Error,* p. 101.

15. *Epperson v. Arkansas,* 393 U.S. 97 (1968).

16. "Memorandum Opinion," *Hendren,* 19–20, quoted in Larson, *Trial and Error,* p. 145.

17. Institute for Creation Research, "The Institute for Creation Research FAQs," http://www.icr.org/home/faq/ (updated February 26, 2007).

18. Alex Heard, "Creationist Movement Appears to Be Slowed by Loss in Arkansas," *Education Week,* February 17, 1982.

19. Brenda Tirey, "Senate Approves Bill to Distribute Tax on Premiums to Fire Fighter Pension Fund," *Arkansas Gazette,* March 13, 1981, p. 10-A, quoted in Larson, *Trial and Error,* p. 151.

20. I will have more to say about these ideas in a later chapter when I look at the conceptual rather than legal history of creationism, but for now suffice it to say that this description of the creation position represented the thinking of the creationist movement.

21. Marcel La Follette, "Creationism in the News: Mass Media Coverage of the Arkansas Trial," in Marcel La Follette, *Creationism, Science, and the Law: The Arkansas Case* (Cambridge, MA: MIT Press, 1983), p. 194.

22. Norman Geisler, *The Creator in the Courtroom: "Scopes II"* (Milford: Mott Media, 1982), p. 208.

23. Langdon Gilkey, *Creationism on Trial: God and Science at Little Rock* (Charlottesville: University Press of Virginia, 1985), p. 104. Gilkey notes in his account that "Reading over that record of my own testimony, competently as it was done considering the difficulties of reporting and then transcribing often complex and frequently esoteric matters, I realize how much of what was actually said is inevitably omitted from such an official record" (p. 244).

24. Gilkey, *Creationism on Trial,* p. 76.

25. Brig Klyce, "Chandra Wickramasinghe in Arkansas," http://www.panspermia.org/chandra.htm (accessed February 26, 2007).

26. Gilkey, *Creationism on Trial,* p. 141.

27. Robert Gentry and Donald Chittick.

28. *McLean v. Arkansas Board of Education,* Judge William Overton (U.S. Dis. Ct. 1982), quoted in Michael Ruse, ed., *But Is It Science? The Philosophical Question in the Creation/Evolution Controversy* (Amherst, NY: Prometheus, 1996), p. 326.

29. "Creationism in Schools: The Decision in *McLean Versus the Arkansas Board of Education,*" *Science* 215, no. 4535 (1982): 934–43.

30. Duane Gish, Introduction, in Geisler, *Creator in the Courtroom,* p. 1.

31. Geisler, *Creator in the Courtroom,* p. x.

32. J. P. Moreland, *Christianity and the Nature of Science: A Philosophical Investigation* (Grand Rapids, MI: Baker, 1989).

33. Larry Laudan, "Science at the Bar—Cause for Concern," in Ruse, ed., *But Is It Science?* pp. 351–55.

34. *Edwards v. Aguillard* 482 U.S. 578 (5th Cir. Ct. App. 1987).

35. Affidavit of Dr. Dean H. Kenyon, filed in *Edwards v. Aguillard* 482 U.S. 578 (5th Cir. Ct. App. 1987), http://www.talkorigins.org/faqs/edwards-v-aguillard/kenyon.html (accessed February 16, 2007).

36. Joel Cracraft, "Reflections on the Arkansas Creation Trial," *Paleobiology* 8, no. 2 (Spring 1982): 83–89.

37. Stephen Jay Gould, "Evolution's Erratic Pace," *Natural History* 86, no. 5 (May 1977): 14.

38. Larson, *Trial and Error,* p. 177.

39. Wendell Bird, *The Origin of Species Revisited* (Nashville, TN: Philosophical Library, 1991), p. 447.

40. Joan Biskupic, "Justice Brennan, Voice of Court's Social Revolution, Dies," *Washington Post,* July 25, 1997.

41. *Edwards v. Aguillard,* 482 U.S. 578 (1987).

42. One curious feature of twentieth-century creationism, which relies heavily on flood geology, is the nearly total absence of Ph.D. geologists from the ranks of the creationists. Ronald Numbers reports on the challenges the creationists have faced in trying to get a geology student to complete a doctorate without abandoning the central tenets of creationism. See "Flood Geology Without Flood Geologists," in *The Creationists: From Scientific Creationism to Intelligent Design,* expanded ed. (Cambridge, MA: Harvard University Press, 2006), pp. 301–11.

43. Though I outline the conceptual origins of this movement in a later chapter, this is a good place to discuss the legal implications of this new face of creationism. I also note here that some scholars, including the nonpartisan historian Jon Roberts, whom I respect enormously, disagree with me that ID is simply a new face on the old creationism. My equation of the two derives from my conviction that both of them are, when all is said and done, really little more than a catalog of minor challenges to evolutionary theory.

44. Percival Davis and Dean H. Kenyon, *Of Pandas and People,* ed. Charles B. Thaxton (Dallas: Haughton, 1993), pp. 99–100.

45. Burt Humburg and Ed Brayton, "Kitzmiller et al. versus Dover Area School Disctrict," *eSkeptic,* December 20, 2005, http://www.skeptic.com/eskeptic/05-12-20.html (accessed February 26, 2007).

46. Peter Slevin, "Teachers, Scientists Vow to Fight Challenge to Evolution," *Washington Post,* May 5, 2005, p. A03.

47. Nick Matzke, "*Of Pandas and People,* the Foundational Work of the 'Intelligent Design' Movement," National Center for Science Education, November 23, 2004, http://www.ncseweb.org/resources/articles/8442_1_introduction_iof_pandas__11_23_2004.asp.

48. Humburg and Brayton, "Kitzmiller et al. versus Dover Area School District."

49. Davis and Kenyon, *Of Pandas and People,* pp. 99–100.

50. Dean Kenyon, Foreword, in Henry Morris and Gary Parker, *What Is Creation Science?* (Green Forest, AR: Master Books, 1997), p. 2.

51. Wayne Frair and Percival Davis, *A Case for Creation,* 3d ed. (Kansas City, MO: Creation Research Society Books, 1983).

52. Lauri Lebo, "In the Judge's Hands," *York Daily Record/Sunday News Online,* November 5, 2005, http://www.ydr.com/search/ci_3219243 (February 26, 2007). Some board members testified the reporters made up the quotations.

53. *Kitzmiller v. Dover,* case no. 04cv2688132, 132 (Pa. District Ct. 2005).

54. *Kitzmiller v. Dover,* case no. 04cv2688132, 137 (Pa. District Ct. 2005).

55. *Kitzmiller v. Dover,* case no. 04cv2688132, 82 (Pa. District Ct. 2005).

56. *Kitzmiller v. Dover,* case no. 04cv2688132, 88 (Pa. District Ct. 2005).

57. "Weblog: Dover Board Lied! Intelligent Design Died!" *Christianity Today,* http://ctlibrary.com/34400 (accessed March 4, 2007).

58. *Kitzmiller v. Dover,* case no. 04cv2688132, 49 (Pa. District Ct. 2005).

59. *Kitzmiller v. Dover,* case no. 04cv2688132, 138 (Pa. District Ct. 2005).

60. David Dewolf et al., *Traipsing into Evolution* (Seattle: Discovery Institute, 2006).

61. Clarence Darrow, *Attorney for the Damned,* ed. Arthur Weinberg (New York: Simon & Shuster, 1989), pp. 217–18.

62. Irving Stone, *Clarence Darrow for the Defense* (New York: Doubleday, 1941), p. 498.

63. Gilkey, *Creationism on Trial,* p. 93.

64. Poll: Majority Reject Evolution, *CBS News,* October 23, 2005, http://www.cbsnews.com/stories/2005/10/22/opinion/polls/main965223.shtml.

CHAPTER 5: **THE EMPEROR'S NEW SCIENCE**

1. Jennifer Harper, "Americans Still Hold Faith in Divine Creation," *Washington Times,* June 9, 2006, http://washingtontimes.com/national/20060608-111826-4947r.htm (accessed February 10, 2007). What is not clear from the poll is exactly how many Americans might still hold to the gap theory or some other creation model in which the earth itself might be much older than the human race. My experience suggests, however, that most creationists are young-earth creationists.

2. A typical example is Robert Snyder, Barbara L. Mann, et al., *Earth Science: The Challenge of Discovery* (Lexington, MA: Heath, 1991), p. 377.

3. Most people think winter is colder than summer because the earth is farther from the sun during the winter, a particularly dumb mistake, considering that the Southern Hemisphere experiences summer while the Northern Hemisphere is in winter.

4. Ellen Gould White, *Patriarchs and Prophets* (Hagerstown, MD: Review and Herald Publishing, 1958), pp. 107–8.

5. George McCready Price, *The New Geology* (Mountain View, CA: Pacific Press, 1923).

6. Price, *New Geology,* pp. 676–77.

7. Price, *New Geology,* pp. 627–29.

8. Martin Gardner, *In the Name of Science* (New York: Putnam, 1952), p. 129.

9. Price, *New Geology,* p. 652. A preflood paradisial climate, which he called the "eternal spring," was a staple in Price's writings, inferred from fossils of some non-Arctic organisms found at the poles. Geological evidence for prehistoric ice at the poles is dismissed as ambiguous. See his *Evolutionary Geology and the New Catastrophism* (Mountain View, CA: Pacific Press, 1926), pp. 258–61. The same argument also appears in his *The Fundamentals of Geology* (Mountain View, CA: Pacific Press, 1913), pp. 195–98.

10. Price, *New Geology,* pp. 655–56, 706.

11. Price, *New Geology,* p. 661.

12. Price, *New Geology,* p. 594.

13. George McCready Price, "Letter to the Editor of *Science* from the Principle Scientific Authority of the Fundamentalists," *Science* 63, no. 1627 (March 5, 1926): 259.

14. Stephen Leacock, *Sunshine Sketches of a Little Town* (UK: Echo Library, 2006), p.71.

15. George McCready Price, *Illogical Geology: The Weakest Point in the Evolutionary Theory* (Los Angeles: Modern Heretic Company, 1906).

16. The most influential of these authorities was perhaps Bernard Ramm, a leading theologian and author of many books on Christian theology. His *The Christian View of Science and Scripture* (Grand Rapids, MI: Eerdmans, 1954) was widely viewed as a definitive. It argued that Christians should accept both the great age of the earth and the nonuniversality of Noah's flood on both biblical and scientific grounds.

17. Ronald Numbers, *The Creationists: From Scientific Creationism to Intelligent Design,* expanded ed. (Cambridge, MA: Harvard University Press, 2006), p. 218.

18. Rice University, as it is now known, was founded by William Marsh Rice in 1912 with an endowment so generous that the university did not have to charge tuition until 1965.

19. Ramm, *Christian View of Science and Scripture,* p. 117.

20. Ramm, *Christian View of Science and Scripture,* p. 28.

21. In his chapter on *The Genesis Flood* in *A History of Modern Creationism* (San Diego: Master Books, 1984), Henry Morris writes: "The Lord marvelously led in the necessary research and writing of the book. Time and again, after encountering a difficult geological (or other) problem, I would pray about it, and then a reasonable solution would somehow quickly come to mind" (p. 153).

22. Morris, *History of Modern Creationism,* p. 150.

23. Numbers, *The Creationists,* pp. 198–99.

24. Price, *New Geology,* p. 706.

25. The index to *The Genesis Flood* has four entries for Price and forty for Ramm, despite the fact that the book draws more heavily on Price than Ramm. Ramm's ideas, of course, are being critiqued, while Price's are being repackaged.

26. Ramm, *Christian View of Science and Scripture,* p. 180.

27. John C. Whitcomb and Henry M. Morris, *The Genesis Flood: The Biblical Record and Its Scientific Implications* (Phillipsburg, NJ: Presbyterian and Reformed Publishing, 1961), p. xx.

28. Ramm, *Christian View of Science and Scripture,* pp. 41–42.

29. Romans 1:22 (ĸJV), obtained at http://bible.cc/romans/1-22.htm.

30. Isaac Asimov, "Is Big Brother Watching?" *The Humanist* 44.4 (July/August 1984): 6–10, 33.

31. Whitcomb and Morris, *Genesis Flood,* p. 222.

32. Daisie Radner and Michael Radner, *Science and Unreason* (Belmont, CA: Wadsworth, 1982). This book looks at pseudoscience and fringe ideas using case studies on flat-earth cosmology, ancient astronauts, biorhythms, creationism, Velikovsky, the bicameral mind, and parapsychology.

33. Michael Shermer, *Why People Believe Weird Things: Pseudoscience, Superstition, and Other Confusions of Our Time* (New York: Holt, 2002).

34. "The Monkey Suit," *The Simpsons,* FOX, May 14, 2006; "The Untitled Griffin Family History," *Family Guy,* FOX, May 14, 2006. On *Saturday Night Live,* NBC, October 29, 2005, Tina Fey read the following "news" item: "The latest Gallup poll found that 66 percent of Americans think President Bush is doing a poor job in Iraq. The remaining 34 percent believe that Adam and Eve rode to church on dinosaurs."

35. Larry Vardiman, "Sensitivity Studies on Vapor Canopy Temperature Profiles," Proceedings of the 4th International Conference on Creationism (Pittsburg: Creation Science Fellowship, 1998), pp. 607–18, 616.

36. Don DeYoung, *Thousands, Not Billions: Challenging an Icon of Evolution: Questioning the Age of the Earth* (Green Forest, AR: Master Books, 2005), p. 60.

37. Whitcomb and Morris, *The Genesis Flood,* p. 167.

38. Richard Dawkins, *The Blind Watchmaker* (New York: Norton, 1986); Stephen Jay Gould, *Wonderful Life: The Burgess Shale and the Nature of History* (New York: Norton, 1990); Ken Miller, *Finding Darwin's God* (London: Harper Perennial, 2000); Darrel Falk, *Coming to Peace with Science* (Downers Grove, IL: InterVarsity, 2004).

39. Richard Dawkins, "Put Your Money on Evolution," *New York Times,* April 9, 1989, accessed via http://proquest.umi.com.

40. Daniel Dennett, *Darwin's Dangerous Idea: Evolution and the Meaning of Life* (New York: Simon & Schuster, 1995) p. 519.

41. Henry Morris, *The Long War Against God: The History and Impact of the Creation/Evolution Controversy* (Green Forest, AR: Master Books, 2000), p. 260.

42. Transnational Association of Christian Colleges and Schools, "Accreditation Standards," http://www.tracs.org/standards.htm (accessed March 1, 2007).

43. Numbers, *The Creationists,* pp. 399–431.

CHAPTER 6: CREATIONISM EVOLVES INTO INTELLIGENT DESIGN

1. I am excluding Stephen Meyer's controversial 2004 publication in *Proceedings of the Biological Society of Washington,* a peer-reviewed scientific journal. The paper, titled "The Origin of Biological Information and the Higher Taxonomic Categories" (117 [2]: 213–39), was later withdrawn by the publisher. The editor of *Proceedings,* a known supporter

of intelligent design, was accused of going outside the usual review procedures in order to get Meyer's paper published. The editor was fired. In any case, Meyer was not advocating flood geology; he is, in fact, a historian and philosopher of science, not a scientist.

2. Rank-and-file creationists typically did not read the "secular" literature critiquing their position, and unscrupulous polemicists would often keep repeating claims, even after their fellow creationists had withdrawn them. This odd state of affairs has led to the publication of books like Mark Isaak's *The Counter-Creation Handbook* (Berkeley: University of California Press, 2007), which simply lists all the standard claims of the creationists with their standard refutation. There are hundreds of claims organized under the headings of philosophy, biology, paleontology, geology, astronomy and cosmology, physics and mathematics, biblical creationism, and intelligent design. Most of the claims continue to appear with regularity in popular anti-evolutionary books.

3. See, for example, Davis Young, *Christianity and the Age of the Earth* (Grand Rapids, MI: Zondervan, 1982). At the time he wrote the book, Young was a fully credentialed professor of geology at Calvin College, one of America's leading liberal arts colleges, and strongly in the evangelical tradition. Young's credentials, affiliation, and strong public Christian faith should have made him an authority on this topic, but his book fell stillborn from the press and was ignored by young-earth creationists.

4. Robert Schadewald, "The 1998 International Conference on Creationism," *NCSE Reports*, www.ncseweb.org/resources/rncse_content/vol18/9954_the_1998_international_confere_12_30_1899.asp (accessed February 13, 2007). Wise's ruthless honesty is shared by fellow creationists Paul Nelson and John Mark Reynolds. See Paul Nelson and John Mark Reynolds, "Young Earth Creationism," in Paul Nelson, Robert C. Newman, and Howard J. Van Till, eds., *Three Views on Creation and Evolution* (Grand Rapids, MI: Zondervan, 1999), pp. 41–75.

5. Nelson and Reynolds, "Young Earth Creationism," p. 51.

6. Ken Miller, *Finding Darwin's God* (London: Harper Perennial, 2000); Howard Van Till, *The Fourth Day* (Grand Rapids, MI: Eerdmans, 1986); Davis Young, *Christianity and the Age of the Earth* (Thousand Oaks, CA: Artisan, 1988); Darrel Falk, *Coming to Peace with Science* (Downers Grove, IL: InterVarsity, 2004); Karl Giberson, *Worlds Apart: The Unholy War Between Science and Religion* (Kansas City: Beacon Hill Press of Kansas City, 1993).

7. M. Conrad Hyers, *The Meaning of Creation: Genesis and Modern Science* (Louisville, KY: Westminster/John Knox, 1984); John Haught, *God After Darwin* (Boulder, CO: Westview, 2000).

8. Taken from a sermon at Coral Ridge Presbyterian Church, http://www.leaderu.com/issues/fabric/chap05.html (accessed February 17, 2007).

9. Glen Kuban, "A Matter of Degree: An Examination of Carl Baugh's Alleged Credentials," *NCSE Reports* 9, no. 6 (November–December 1989), http://paleo.cc/paluxy/degrees.htm (accessed February 17, 2007).

10. Carl E. Baugh, *Panorama of Creation* (Oklahoma City, OK: Creation Evidences Museum, 1989), pp. 49, 51.

11. http://www.tbn.org/index.php/2/4/p/3.html (accessed February 17, 2007).

12. http://www.kent-hovind.com (accessed February 17, 2007).

13. Henry Morris, *The Twilight of Evolution* (Philadelphia: Presbyterian and Reformed Publishing, 1963).

14. Tom McIver's encyclopedic *Anti-Evolution: A Reader's Guide to Writings Before and After Darwin* (Baltimore: Johns Hopkins University Press, 1988) lists many titles announcing the end of evolution. There are three alone titled *The Collapse of Evolution,* the most recent and substantial of which is Scott Huse, *The Collapse of Evolution* (Grand Rapids, MI: Baker, 1997). For an Islamic perspective, but one drawing heavily on Christian creationism, see Harun Yahya, *The Evolution Deceit: The Scientific Collapse of Darwinism and Its Ideological Background* (New York: Global, 2001).

15. "The Untitled Griffin Family History," *Family Guy,* FOX, May 14, 2006.

16. Stephen Jay Gould, "Impeaching a Self-Appointed Judge," *Scientific American* 267 (July 1992): 118–21.

17. Phillip Johnson, *Darwin on Trial* (Washington, DC: Regnery Gateway, 1991), p. 154.

18. William Dembski. *Mere Creation: Science, Faith, and Intelligent Design* (Downers Grove, IL: InterVarsity, 1998).

19. Thomas Woodward, *Doubts About Darwin: A History of ID* (Grand Rapids, MI: Baker, 2003); see also his *Darwin Strikes Back: Defending the Science of ID* (Grand Rapids, MI: Baker, 2006).

20. K. W. Giberson and D. A. Yerxa, *Species of Origins: America's Search for a Creation Story* (New York: Rowman & Littlefield, 2002), pp. 193–233.

21. Barbara Forrest and Paul R. Gross, *Creationism's Trojan Horse: The Wedge of Intelligent Design* (Oxford: Oxford University Press, 2004); Robert Pennock, *Tower of Babel: The Evidence Against the New Creationism* (Cambridge, MA: MIT Press, 1999).

22. Ahmanson has served for years on the board of R. J. Rushdoony's Chalcedon Foundation, a "Christian Reconstructionist" organization seeking to impose Old Testament law on America. Reconstructionists, to take one example, propose capital punishment for "crimes" like blasphemy, heresy, adultery, and homosexuality. See Forrest and Gross, *Creationism's Trojan Horse,* pp. 264–67.

23. William Dembski, *Intelligent Design: The Bridge Between Science and Theology* (Downers Grove, IL: InterVarsity, 1999).

24. William Dembski, *The Design Revolution: Answering the Toughest Questions About Intelligent Design* (Downers Grove, IL: InterVarsity, 2004), p. 45.

25. Michael Behe, "The Modern Intelligent Design Hypothesis," *Philosophia Christi,* series 2, vol. 3, no. 1 (2001): 165.

CHAPTER 7: **HOW TO BE STUPID, WICKED, AND INSANE**

1. Cornelia Dean, "Evolution Takes a Back Seat in U.S. Classes," *New York Times,* February 1, 2005.

2. Cedarville College, for example, publishes the following statement on its college Web site: "We believe that the universe, solar system, earth, and life were all created recently by an omnipotent, omniscient God during six literal 24 hour days, as described in Genesis Chapters 1 and 2"; see http://www.cedarville.edu/academics/sciencemath/origins.htm (accessed May 17, 2007). Cedarville was accredited in 1975 and is consistently ranked in the top tier of Midwest colleges by *U.S. News & World Report.* Liberty University proudly displays its accreditation by the anti-evolutionary Transnational Association of Christian

Colleges and Schools (TRACS) above its conventional accreditation by the Southern Association of Colleges and Schools; see http://www.liberty.edu/index.cfm?PID=284 (accessed May 18, 2007).

3. Daniel Dennett, *Darwin's Dangerous Idea: Evolution and the Meaning of Life* (New York: Simon & Schuster, 1995); Richard Dawkins, *The God Delusion* (Boston: Houghton Mifflin, 2006); Phillip Johnson, *Defeating Darwin by Opening Minds* (Downers Grove, IL: InterVarsity, 1997); Henry Morris, *The Long War Against God: The History and Impact of the Creation/Evolution Controversy* (Green Forest, AR: Master Books, 2000).

4. Ken Miller, *Finding Darwin's God* (London: Harper Perennial, 2000); John Haught, *God After Darwin* (Boulder, CO: Westview, 2000); Darrel Falk, *Coming to Peace with Science* (Downers Grove, IL: InterVarsity, 2004); Francis Collins, *The Language of God: A Scientist Presents Evidence for Belief* (New York: Free Press, 2006).

5. *Evolution: International Journal of Organic Evolution.*

6. *Answers; Origins & Design; Creation Ex Nihilo; Creation Research Society Quarterly.*

7. *NCSE Reports.*

8. *Perspectives on Science & Faith.*

9. "Statement on Teaching Evolution," National Association of Biology Teachers, March 15, 1995, http://darwin.eeb.uconn.edu/Documents/NABT.htm (accessed March 30, 2007).

10. Simon Conway Morris, *Life's Solution: Inevitable Humans in a Lonely Universe* (Cambridge: Cambridge University Press, 2003), p. 282.

11. Conway Morris, *Life's Solution,* pp. 284–85.

12. Robert Wright, *Non-Zero: The Logic of Human Destiny* (New York: Pantheon, 2000).

13. Phillip Johnson, *Objections Sustained: Subversive Essays on Evolution, Law and Culture* (Downers Grove, IL: InterVarsity, 1995), p. 86.

14. Johnson, *Objections Sustained,* p. 91.

15. Louise Lamphere et al., *August 2001 Joint Letter from Scientific and Educational Leaders on Evolution in H.R.1.,* Letter to House Committee on Education and the Workforce, October 9, 2001, AGI Government Affairs Program, http://www.agiweb.org/gap/legis107/evolutionletter.html (accessed March 6, 2007).

16. My relativity professor at Rice University, where I did my doctoral work in the early 1980s, used Einstein's book *The Meaning of Relativity,* written in 1922. The text was entirely adequate and pedagogically superior to many more recent treatments. A 1922 textbook on evolution, by contrast, would be hopelessly out-of-date if used today, even though, in 1922, evolution was decades older than relativity.

17. Stephen Jay Gould, *The Structure of Evolutionary Theory* (Cambridge, MA: Harvard University Press, Belknap Press, 2002); Richard Dawkins, *The Ancestor's Tale: A Pilgrimage to the Dawn of Evolution* (Boston: Houghton Mifflin, 2004).

18. Conway Morris, *Life's Solution,* pp. 4–5.

19. For an excellent survey of Galileo propaganda, see William R. Shea and Mariano Artigas, *Galileo Observed: Science and the Politics of Belief* (Sagamore Beach, MA: Science History Publications, 2006).

20. Richard Dawkins, review of Donald Johanson and Maitland Edey, *Blueprints: Solving the Mystery of Evolution, New York Times,* April 9, 1989, sec 7.

21. Dawkins, *The Ancestor's Tale,* pp. 613–14.

22. Carl Sagan, *Cosmos* (New York: Ballantine, 1985), p. 345.

23. Stephen Jay Gould, *Wonderful Life: The Burgess Shale and the Nature of History* (New York: Norton, 1990), p. 323.

24. Edward O. Wilson, *On Human Nature* (Cambridge, MA: Harvard University Press, 1978), p. 201.

25. Steven Weinberg, *The First Three Minutes: A Modern View of the Origin of the Universe,* updated ed. (New York: Basic Books, 1993), pp. 154–55.

26. Richard Dawkins, *The Blind Watchmaker* (New York: Norton, 1986), p. 6.

27. Morris, *The Long War Against God,* p. 25.

28. Morris, *The Long War Against God,* p. 327.

29. Jonathan Sarfati, *Refuting Evolution: A Response to the National Academy of Sciences' Teaching About Evolution and the Nature of Science* (Green Forest, AR: Master Books, 1999), p. 16.

30. Phillip Johnson, *Reason in the Balance: The Case Against Naturalism in Science, Law and Education* (Downers Grove, IL: InterVarsity, 1995), p. 12.

31. Richard Lewontin, "Billions and Billions of Demons," *New York Review,* January 9, 1997, p. 31 (emphases in original).

32. Richard E. Dickerson, "Random Walking," *Journal of Molecular Evolution* 34 (April 1992): 277.

33. William B. Provine, "Influence of Darwin's Ideas on the Study of Evolution," *BioScience* 32 (June 1982): 506.

34. Randy Thornhill and Craig T. Palmer, *A Natural History of Rape: Biological Bases of Sexual Coercion* (Cambridge, MA: MIT Press, 2000).

CHAPTER 8: EVOLUTION AND PHYSICS ENVY

1. William R. Overton, United States District Court Opinion, *McLean v. Arkansas Board of Education,* in Michael Ruse, ed., *But Is It Science? The Philosophical Question in the Creation/Evolution Controversy* (Amherst, NY: Prometheus, 1996), pp. 307–31.

2. Phillip Johnson, *Darwin on Trial* (Washington, DC: Regnery Gateway, 1991), pp. 145–54; Henry M. Morris, *Scientific Creationism* (Green Forest, AR: Master Books, 1985), pp. 6–7.

3. "I have changed my mind about the testability and logical status of the theory of natural selection; and I am glad to have an opportunity to make a recantation." "Natural Selection and the Emergence of Mind," *Dialectica* 32, no. 3–4 (1978): 339–55, http://www.geocities.com/criticalrationalist/popperevolution.htm.

4. Both Johnson and Morris (see n. 2) wrote their remarks years after Popper changed his mind about evolution being falsifiable.

5. Francis Crick, *What Mad Pursuit: A Personal View of Scientific Discovery* (New York: Basic Books, 1988), pp. 138–39.

6. Johnson, *Darwin on Trial,* p. 154.

7. Phillip Johnson, *Reason in the Balance: The Case Against Naturalism in Science, Law and Education* (Downers Grove, IL: InterVarsity, 1995), p. 90.

8. Duane Gish, *Evolution: The Fossils Still Say No!* (San Diego, CA: Master Books, 1995), pp. 81–82.

9. Jonathan Wells, *Icons of Evolution: Science or Myth?* (Washington, DC: Regnery, 2000), p. 8.

10. Johnson, *Darwin on Trial,* p. 13. Johnson has such a poor grasp of science that he can't see the illogic in his claim that "a scientist outside his field of expertise is just another layman" (p. 13). Does he really believe that a chemist or physicist is no more qualified to speak about biology than a bricklayer or novelist?

11. Ken Ham, *The Lie: Evolution* (El Cajon, CA: Creation-Life Publishers, 1987), p. 21.

12. Niles Eldredge, *Darwin: Discovering the Tree of Life* (New York: Norton, 2005), p. 228.

13. Ernst Mayr, *What Evolution Is* (New York: Basic Books, 2001), p. 21.

14. Mayr, *What Evolution Is,* p. 31.

15. Charles Darwin, *On the Origin of Species by Means of Natural Selection* (New York: Avenel, 1979), p. 422.

16. Richard Dawkins, *River Out of Eden: A Darwinian View of Life* (New York: Basic Books, 1995), p. 78.

17. Darrel Falk, *Coming to Peace with Science* (Downers Grove, IL: InterVarsity, 2004), pp. 171–98.

18. M. Shimanura et al., "Molecular Evidence from Retroposons That Whales Form a Clade Within Even-Toed Ungulates," *Nature* 388 (1997): 666–68. The authors note that their conclusions confirm data from paleontological, morphological, and molecular studies (p. 666).

19. Michael J. Behe, *The Edge of Evolution: The Search for the Limits of Darwinism* (New York: Free Press, 2007).

20. J. P. Moreland, ed., *The Creation Hypothesis: Scientific Evidence for an Intelligent Designer* (Downers Grove, IL: InterVarsity, 1994).

21. Arthur Peacocke, *Evolution: The Disguised Friend of Faith?* (Philadelphia: Templeton Foundation Press, 2004)

CONCLUSION: PILGRIM'S PROGRESS

1. E. O. Wilson, *Biophilia* (Cambridge, MA: Harvard University Press, 1984).

2. This view of God's creative involvement does nothing, in principle, to compromise the traditional biblical affirmation that God "numbers the hairs on our heads" or "cares about the sparrow's fall." There is no reason why God cannot be intimately concerned about such details, just because God is the not the immediate creative agent behind those details.

3. Stephen Jay Gould, *Full House: The Spread of Excellence from Plato to Darwin* (New York: Harmony, 1996), p. 216.

4. Mariano Artigas, Thomas F. Glick, and Rafael A. Martinez, *Negotiating Darwin: The Vatican Confronts Evolution, 1877–1902* (Baltimore: Johns Hopkins University Press, 2006); David N. Livingstone, *Darwin's Forgotten Defenders: The Encounter Between Evangelical Theology and Evolutionary Thought* (Grand Rapids, MI: Eerdmans, 1987).

5. Freeman Dyson, *Disturbing the Universe* (New York: Basic Books, 1979), p. 250.

INDEX

Abbott, Lyman, 59–60
ACLU (American Civil Liberties Union), 16, 88–89, 97, 98, 100, 104–7, 112, 113, 116
Adam and Eve, 4, 6–13, 52, 54–55, 57, 215
Agassiz, Louis, 48, 49
age of earth: as billions of years, 204, 205; day-age theory, 49–52, 129, 131, 136, 137, 195; gap theory, 49, 51–52, 131; 19th-century views on, 48, 49; reconciling with Genesis, 5–8, 134–35, 225n5
agnosticism, ix, 28, 38, 213
Aguillard, Daniel, 104, 105, 113
altruism, 12, 174
Ancestor's Tale (Dawkins), 43, 177, 211
Anglican Church, 21–22
animals: as adaptive, 36, 55; cruelty amongst, 32–35, 37–38, 40, 161–62, 210; geographical distribution of, 33–34, 49; human kinship with, 13–15; kindness amongst, 13–14, 209
Ankerberg, John, 19
Answers in Genesis, 20, 21, 144
anti-evolution groups: appeal to common-sense, 119, 128, 159; differences separating, 205; rhetoric engulfing, 141; template for work of, 128. See also creationism; intelligent design (ID) movement; young-earth creationism
anti-evolution laws, 88, 94–96, 98
anti-Semitism, 77–78
apocalypticism, 58, 59
Apple, Inc., 63–65, 80, 81
Arkansas Act 590, 98–103, 108

Arkansas trial (Scopes II), 16, 98–103, 111, 117, 119, 186
Arnold, Matthew, 45–46
Asimov, Issac, 138, 140
astronomy, 5, 24–26, 158–60
atheism, 40, 43, 44–46, 110, 180, 183
Atkins, Peter, 44, 175, 183
Augustine of Hippo, Saint, 53
Ayala, Francisco, 101, 102

Bad design, 162–63, 210
Balanced Treatment Act, 103–9
Baltimore Sun (newspaper), 86, 89–90
Baugh, Carl, 147, 148, 153
Beagle voyage, 22–23, 26, 27, 29, 31–33, 35, 38, 41, 49, 164
Behe, Michael, 54, 111, 115, 164, 183, 205
Bell Curve (Herrnstein/Murray), 79
Bible: authorship of, 3–4, 29; conflicts around interpretation, 135–36; contemporary biblical scholarship, 8–9, 15–16, 119, 146–47, 149; faulty translations, 8, 49–50; Scofield, 51–52; Strauss's critical analysis, 44–47, 59; symbolic reading of, 5, 9, 52, 53. See also Genesis creation account; literalism; Noah's flood
big bang theory, 5, 54, 191
biochemistry, 194, 200–201, 203
biogeography, 33–34, 54, 189, 190, 194, 198–99, 204
biologists, 53–54, 193–94
Biology: A Search for Order in Complexity, 97
Biology for Beginners (Moon), 93

biophilia, 210
Bird, Wendell, 97–98, 104–11, 116, 152
Black, Hugo, 96
blacks, 52, 76, 77
Blind Watchmaker (Dawkins), 43, 179
blood-clotting mechanism, 156, 157, 160, 161, 163, 180
bonobo apes, 13–14, 209
Book of Nature, 2, 3
Bowler, Peter, 57
Brahe, Tycho, 26, 32
Bryan, William Jennings, 50, 61, 82, 83, 86–94, 96–99, 109, 114, 117, 118, 124, 126, 127, 129, 130, 133, 173, 183
Buck, Carrie, 61, 74
Buckingham, William, 114, 117
Buckland, William, 51
Buffon, Georges-Louis Leclerc de, 50, 53

Can a Darwinian Be a Christian? (Ruse), 16
capitalism, 63–65, 68, 80, 83, 173, 183
carbon dating, 5, 122, 127, 139, 142
Carnegie, Andrew, 65, 80–81, 173, 183
cartoons, Christian, 139, 144, 145
Catholic Church, 15, 28, 74, 172, 192
cats, 32, 35, 38, 40
CBS News polls, 6, 118, 225n4
Cedarville College, 165, 237n2
cellular reproduction, 191, 217–18
Chalmers, Thomas, 51
chance, 211, 219, 220
Chesterton, G. K., 12
Christian Heritage College, 139
Christian intellectuals, 119, 135, 146–47, 149, 153
Christian View of Science and Scripture (Ramm), 131, 135–37
Christianity. See religion
Civic Biology (Hunter), 73–74, 89
Clark, Steven, 100
Cold War, 94, 130
Collins, Francis, vii-x, 15
common descent: development of new species and, 192; and embryology studies, 200–201; as fact, 53–54, 204; and the fossil record, 54, 197–98, 204; genetics supporting, viii–ix, 189, 201–4; and the origin of life, 54, 217–19; and uniqueness of humankind, 13–14

common structures, 199–200
commonsense, 119, 128, 159
comparative anatomy/physiology, 54, 190, 194, 199–200
comparative biochemistry, 194, 200–201, 203
comparative DNA, 54
competition, 14, 33, 36, 63–65, 67, 68
contemporary biblical scholarship, 9, 15–16, 119, 146–47, 149
Conway Morris, Simon, 167–68, 171
cosmic evolution, 65–66, 174, 191, 218
Cosmos (Sagan), 177
Coulter, Ann, 77, 83, 153
Creation (Wilson), 174, 178
creation account. See creationism; Genesis creation account
Creation Museum, vii, 27
creation museums, vii, 21, 139, 144, 149, 195
creationism: arguing evolution is not science, 186–89; assaults on Darwin, 19–21, 27, 147; as pseudoscience, 139, 140, 147–48; based on Genesis creation account, 4–6, 151–52; and biogeography, 204; birth of, 1–2, 124, 130–32, 138, 142, 145; "brand-name," 138–41; day-age theory, 49–52, 129, 131, 136, 137, 195; failure as a science, 145–49, 210; gap theory, 49, 51–52, 131; lack of unity in, 205; literature of, 145–46, 148; no accounting for bad design, 162–63, 210; old-earth, ix, 132, 133, 205; as oversimplifying models of origin, 102, 107–8, 138; popularity of, 6, 17, 118, 141–42; pre-Darwinian theories of, 48–53; progressive creationism, 60, 132; research centers to support, 138, 139, 141, 143–45; satire of, 143, 148; "secularized," 97–98, 104, 107. See also intelligent design (ID) movement; young-earth creationism
Crick, Francis, 44, 175, 188
culture war: contentious nature of, 43–44, 165–66; creating a false dichotomy, 215–16; as "elites" versus "ordinary taxpayers," 91–92; evolution as "dissolving" religion, 9–11, 16–17, 82, 149, 215; evolution as religion, 174–82, 194; impact on teaching evolution, 165–66; legacy of,

culture war *(continued)*
    172–74, 182–84; literature on the, 166;
    making peace, viii–x, 15–18, 206, 215;
    manufacturing controversy over evolu-
    tion, 58, 91–92, 147; in politics, 169–72;
    worldview as central to, 149, 166–69

Darrow, Clarence, 48, 50, 86, 87, 89–93,
    99, 117
Darwin, Annie, 39–40, 71
Darwin, Charles: agnosticism of, 28, 38–
    40; bothered by cruelty in nature, 32, 34–
    35, 37, 38, 40, 161–62; championed by
    Dawkins, 43, 183; Christian faith of, 17,
    19–20, 23, 26–28, 40–41; connecting
    humans with nature, 208–9, 214, 215;
    creationism's assaults on, 19–21, 27, 147;
    development of his theory, 32–37, 53;
    discussed in The Fundamentals, 60; early
    studies and Beagle voyage, 21–23; and
    evolution as religion, 179–80, 183; influ-
    ences on, 55, 71, 193, 201; and intelligent
    design, 26–27, 31, 155, 161–64; mytho-
    logy around, 16, 19; natural theology em-
    braced by, 27–31, 163; and paradigm
    shifts, 23–25, 31–32; reception in Ame-
    rica, 46–48, 53, 228n15; struggles over
    his loss of faith, 45, 156, 162, 164, 173.
    See also evolutionDarwin, Emma, 39–40
Darwin on Trial (Johnson), 150–51
Darwin's Black Box (Behe), 54, 115
Darwin's Leap of Faith (Ankerberg/Wel-
    don), 19
Dawkins, Richard, 28, 43–46, 140, 171,
    173–77, 179–80, 182–83, 190, 210–13,
    215–16, 220
Dawkins Delusion (McGrath), 43
Dawkins' God (McGrath), 43
day-age theory, 49–52, 129, 131, 136,
    137, 195
Dayton (TN), 87–89, 91, 117
de Waal, Frans, 13
Dembski, William, 54, 111, 153, 157, 159–
    62, 164, 183
Dennett, Daniel, 9–11, 44, 140, 155, 183,
    215–16, 220
design inferences, 154, 158–59
Desmond, Adrian, 41
developmental similarities, 200–201

devil. See Satan
dinosaurs: in creation cartoons, 144; extinc-
    tion before humans, vii, 195, 196, 204;
    extinction of, 66, 192; supposed exis-
    tence with humans, 125, 139, 140, 142,
    148, 204
Discovery Institute, 28, 114–17, 153, 165
distribution of life, 33–34, 54, 189, 190,
    194, 198–99, 204
DNA studies. See genetics
Dover (PA), 111–12, 117
"Dover Beach" (poem; Arnold), 45–46
Dover trial, 111–19, 153
Dyson, Freeman, 219

Eastern Nazarene College, 1, 2
Eddington, Arthur, 185, 187
education system. See public schools debate
Edwards v. Aguillard, 110
Einstein, Albert, 25, 121, 170, 185, 213
Eldredge, Niles, 82
electromagnetic interactions, 217–18
Eliot, T. S., 221
elitism, 70–73, 119
embryology studies, 200–201
ends justifying the means, 66–67
England. See Victorian England
Epperson, Susan, 94–96
equal time debate, 93–98, 103–9
Essay on the Principle of Population (Mal-
    thus), 70–71
eugenics, 61, 72–80, 82, 83, 86, 173, 183
Eugenics Record Office, 73
Europe, 21, 29, 30, 44, 45, 142
evangelical Christians: acceptance of
    evolution, 15, 47, 129, 138; as anti-
    evolutionists, 21, 58; embracing ID/
    young-earth creationism, 58, 114, 130–
    31, 137, 138, 143; fundamentalism's
    negative impact on, 132; the Lady Hope
    myth, 19; subculture of, 144, 145
evidence for theory of evolution: biogeo-
    graphy, 54, 189, 194, 198–99; common
    structures, 199–200; developmental sim-
    ilarities, 200–201; fossil record, viii, 189,
    194–98; genetics, viii–ix, ix, 54, 93, 194,
    201–4; overview on, 189–90, 193–94,
    204–6; overwhelming, viii–ix, 53–54, 93,
    95, 210–11

Evidences of Christianity (Paley), 22
evolution: alternative explanations for, 54–
57; cornerstone of modern biology,
102–3, 183; as descriptive, 64, 69, 80;
development of theory, 32–37, 53; difficulty in defining, 167–69, 176; as dissim-
ilar to physics, 185–89; as inadequate,
152–54, 156–57, 164; modern story of,
171, 174, 189–93; momentum in, 12,
56–57; and natural selection, 36–37, 53–
58, 152–54, 156–57, 188, 200, 211; processes of, 57–58; as pseudoscience, 186,
187, 193; reception by Darwin's contemporaries, viii–x, 41, 46–48; as religion,
174–82, 194; teaching, 17, 165–69, 183,
215; as underdetermined theory, 211–
12; united front supporting, 205–6. See
also common descent; Darwin, Charles;
evidence for theory of evolution
Evolution (television series), 166
Evolution of Christianity (Abbott), 59–60
explanatory filters, 157, 158, 161, 162, 164
extinct species: humans hastening, 67; maladaptive traits of, 54, 56, 66; missing
fossil records of, 188, 196, 212; reconciling with Genesis accounts, 7, 49, 139

Falk, Darrel, 15, 140
falsifiability criterion, 186–87
Falwell, Jerry, 100, 147
Family Guy (television series), 148
feeblemindedness, 61, 73–75, 86, 173
First Amendment rights, 95, 108
First Three Minutes (Weinberg), 174,
178–79
Fitzroy, Robert, 22–23
flood geology. See young-earth creationism
fossil record: gaps in the, 142, 197; and the
geological column, 124–26, 195–96; index fossils, 126; intermediate forms in
the, 107, 187, 197, 212–13; and Noah's
flood, 5, 123, 124, 195, 204; supporting common ancestry, 54, 197–98, 204;
supporting theory of evolution, viii, 189,
194, 195–98
four fundamental interactions, 217, 220
free-market capitalism, 63–65, 68, 80–81,
83, 173, 183
free will, 4, 13

Freud, Sigmund, 151, 186, 187
fundamentalism: attack on evolution, 15,
21, 44, 48, 61, 98; birth of, viii, 46, 59–
62; embracing the day-age/gap theory,
50, 131; in evolution-religion legal battles, 85–87, 91, 100–101, 104, 107; as
hyperorthodox, 38, 132, 134; Morris's
influence on, 1–2, 130–32, 134; reaction to higher criticism, 46, 47; reading
of Genesis story, 4, 8; unwavering support of creationism, 1–3, 109, 128–29,
132, 147, 149
Fundamentals, 60–61, 122, 138

Galapagos Islands, 34, 198
Galileo, viii, 28, 135, 138, 167, 172
Galton, Francis, 72, 75, 79
gap theory, 49, 51–52, 131
Garden of Eden, 8, 10, 52, 134
Geisler, Norman, 100–101, 103
Genesis creation account: Adam and Eve in,
4, 6–13, 52, 54–55, 57, 215; age of earth
and the, 7–8, 49, 225n5; creationism
based on, 4–6, 151–52; and the day-age
theory, 49–52, 131, 136, 137, 195; and
early discussions of Darwin's theory, 48,
49, 53; evolution as "dissolving," 10–
12; and the gap theory, 49, 51–52, 131;
Garden of Eden, 8, 10, 52, 134; literalist
view of, 2, 4, 7–9, 52, 136–38; and original sin, 4, 6–13, 32, 60, 215; as symbolic,
5, 9, 52; and the vapor-canopy hypothesis, 133–34, 139
Genesis flood. See Noah's flood
Genesis Flood (Whitcomb/Morris), viii, 1–
2, 5, 59, 124, 131–40, 142, 145
genetics: and evolution, ix, 54, 93, 189,
190, 194, 201–4; "junk DNA," 163, 203;
pseudogenes, 202–4; seeing God in,
ix, 180
genocide, 61, 76, 78, 173, 174, 183
genomes, viii–ix, 12, 201–3
geological column, 124–26
geology/geologists: on age of earth, 5, 50,
51, 136; on the geological column, 124–
26; on Noah's flood, 33–34, 59, 123,
146; progressive creationism adhering
to, 132; thrust faults, 126–27
Gilkey, Langdon, 100–102

Gish, Duane, 103, 193, 196, 204
God: as author of the Bible, 3–4; as compatible with evolution, ix, 15, 18, 206, 216–17; in creation, 4–5, 209, 210, 216–17; Darwin's questions around, 33–35, 37, 40, 156; evolution theorists on, 173–74, 180, 181; as filling the gaps, 159, 164, 216, 220; in the Garden of Eden, 8, 10, 52; as God of love, 38, 161; in the ID movement, 154–56, 160–62, 216; as incompatible with evolution, 21, 168, 179, 182; limits on knowledge of, 18, 161; in natural history, 22, 27, 164; in Newton's science, 158–59; and 19th-century secularism, 44–46; in progressive creationism, 10, 13, 132; suffering and the existence of, 7, 9, 30, 32, 35, 38–40
God Delusion (Dawkins), 43, 182, 183
"God's Funeral" (poem; Hardy), 46
Gould, Stephen Jay, 101, 107, 140, 146, 151, 165, 167, 171, 174–78, 183, 211, 213
Grand Canyon, 124, 125
gravitational interactions, 217, 218
"Great Disappointment," 58, 122

Haeckel, Ernst, 6, 76
Ham, Ken, 20, 21, 27, 83, 144, 189, 194, 195, 204
Hardy, Thomas, 46
Haught, John, 15
Hawking, Stephen, 175, 176
Heisenberg Uncertainty Principle, 13, 220
hell, 38–39, 60
Hereditary Genius (Galton), 72
"Hereditary Talent and Character" (article; Galton), 72
"Heredity and Variation" (textbook chapter; Hunter), 74
Herrnstein, Richard, 79
hierarchy of values, 67
higher criticism movement, 44–47, 59–60
higher education, 141, 144
Hitler, Adolph, 76, 77, 79, 173
Holmes, Oliver Wendell, 74
Homo sapiens: aberrant behavior in, 80, 83, 173; appearance of, 192–93, 198; blood-clotting mechanism, 156, 157, 160, 161, 163; design problems in, 162–63; dinosaurs as existing with, 125, 139, 140, 142, 148, 204; family tree of, 48–49, 197; impact on other species, 66–67; Lamarck's theories on adaptation, 55; orthogenesis in, 56–57; as unique, 11, 13–15. See also offspring
Hope, Lady (Elizabeth Reid), 19
Hovind, Ken (Dr. Dino), 147–48, 153
humans. See Homo sapiens
Hume, David, 30, 60
Hunter, George William, 74, 89

Ichneumonidae wasp, 34, 35, 37, 38, 40, 161–62
ICR (Institute for Creation Research), 1, 3, 27, 98, 108, 139, 140, 143–45
immigration, 75–76
In Memoriam (Tennyson), 66
index fossils, 126
infanticide, 80, 82, 173, 174, 183
Inherit the Wind (film), 86, 87, 91, 118, 143
intelligence, 79–80
intelligent design: Darwin's early worldview embracing, 26–28, 31, 32; and modern-day ID movement, 31, 163–64; orthogenesis coupled with, 55–57; science's rejection of, ix, 30, 34; tenet of natural theology, 22, 26–31
intelligent design (ID) movement: attack on science, vii, 28, 150–54, 187–89; basic theory, 110, 154–55; and the Dover trial, 111–19, 153; Edwards v. Aguillard case, 110–11; explanatory filter in, 157, 158, 161, 162, 164; failings as a science, ix, 156–60; God in, 216; and historical intelligent design, 31, 163–64; legal strategy behind the, 110–11, 149–50; no accounting for bad design, 162–63; rejecting on theological grounds, 160–62, 216; as science/theology bridge, 54, 160, 162, 164; support for, 17, 113, 152
intermediate forms, 107, 197, 212–13
involuntary sterilization, 61, 73–75, 86, 173
iPod/iTunes, 63, 64, 68, 80, 181
Irish elk, 56

Jesus Christ, viii, 9–11, 39, 45, 47, 59, 60, 155

Jews, 68, 77–78

Johnson, Phillip, 20–21, 28, 44, 110–11, 144, 149–53, 164, 172, 180–83, 189, 193, 194, 205, 216

Jones, John, III, 114–17

Jukes family, 73, 74

"junk DNA," 163, 203

Kennedy, D. James, 83, 147, 153

Kenyon, Dean, 106, 111, 113

Kirchhoff, Alfred, 78

LaHaye, Tim, 139

Lamarck, Jean-Baptiste, 54–55, 57, 65

Language of God (Collins), 15

Le Conte, Joseph, 76–77

Leacock, Stephen, 129

legal battles, 88, 94–96, 98, 117–19. See also public schools debate

leptons, 217, 220

Lewis, C. S., viii, 15

Lewontin, Richard, 181–82

liberalism, viii, 2, 6, 46, 59–61

Liberty University, 145, 165, 237n2

Lie (Ham), 20, 83

Life of Jesus Critically Examined (Strauss), 44–46

Linnaeus, Carolus, 48–49

literalism: Darwin's embrace of, 21–22; Genesis creation story, 2, 4, 7–9, 52, 136–38; practiced by fundamentalists, 3, 8, 109; role in the Morris/Ram debate, 130, 132, 135–36; in young-/old-earth creationism, 205

"Little Gidding" (poem; Eliot), 221

logos (rationality, logic), 155

Long War Against God (Morris), 20, 141

lower classes, 70–75

Luther, Martin, 77

Lyell Charles, 33–34, 49

Malthus, Thomas, 70, 75

mammals, 192, 199–200

mandatory sterilization, 61, 73–75, 86, 173

Many Infallible Proofs (Morris), 1, 3

Marxism, 183, 186–87

materialism, 181–82

mathematics, 70, 187–89, 210, 220

McGrath, Alister, 15–16, 43

McLean v. Arkansas Board of Education, 16, 98–103, 117, 119

meaning of life, ix, 213–14

media, 91, 92, 101, 103, 176

Mencken, H. L., 73, 86–87, 89–90, 92, 99, 118

Mere Creation , 152

Meyer, Stephen, 111, 183

Miethe, Terry L., 107

"might makes right," 68, 183

militarism, 61, 80, 86

Miller, Ken, 15, 140

modernism, 59–61

"Monkey Suit" (Simpson's television episode), 143

"Monkey Trial." See Scopes trial

Monty Python, 109–10, 113

Moon, Truman, 93

Moore, James, 41

morality, 55, 57, 78, 174

More, Thomas, 112

Morris, Henry, viii, 1–3, 5, 6, 20, 27, 59, 83, 98, 124, 130–37, 139–43, 145, 148, 151, 173, 180, 183, 189, 204, 216

Morrow, W. Scott, 106

Most, William G., 107

Murray, Charles, 79

NABT (National Association of Biology Teachers), 167–69, 172, 174, 176, 178, 187, 189

nationalism, 68

Native Americans, 77, 78

natural history, 4, 5, 8–9, 11–15, 125–26, 213, 216–20

Natural History of Rape (Coyne/Berry), 80

natural philosophy, 28, 159

natural selection, 36–37, 53–58, 152–54, 156–57, 188, 200, 211

natural theology, 22, 26–33, 35–37, 163

Natural Theology (Paley), 22, 27, 30, 31

naturalism: acceptance of Lamarck's theory, 55; dawn of, 41; as framework for science, 159–60, 174, 180–81; ID Movement's attack on, 150, 152, 183, 193; as pushing God out, ix, 21, 110, 149–50; reaching beyond traditional bounds of science, 174; teachings on God in nature, 22, 27, 29, 37, 164 nature: apparent cruelty in, 32–35, 37–38, 40, 161–62, 210; competition in, 14, 33, 36; Darwin facilitating connection with, 186, 208–10; as messy, 214–15; mystery in, 221; and natural history, 4, 5, 8–9, 11–15, 125–26, 213, 216–20

Nazism: as diffuser of eugenics, 75, 82; embracing social Darwinism, 61, 68, 76; link to evolution, 61, 65, 82, 83; rationalization for genocide, 75, 76–79, 173, 183, 230n21

Nelson, Paul, 146, 153, 205

New Geology (Price), 124–28, 133

New York Times (newspaper), 91, 185

Newton, Issac, viii, 24, 28–29, 41, 138, 158–60, 167, 185, 210, 214, 220

"No Child Left Behind" program, 169

Noah's flood: and biogeography, 33–34, 37, 49, 198, 204; and extinct species, 7, 139; and the fossil record, 5, 123, 124, 195, 204; Hovind Theory, 148; linked to geological history, 123, 132, 135; as localized event, 123, 129, 131, 135, 146, 205; preflood earth canopies, 127, 133–34, 139, 147; rejected by 19th-century scientists, 33–34, 48; White's narrative, 123–24

nuclear interactions, 217

Of Pandas and People (Davis/Kenyon), 112–16

offspring: and the eugenics movement, 73–75, 79; overproduction and survival, 12, 36, 66, 68, 192; passing desirable traits to, 55, 57, 192

old-earth creationism, ix, 132, 133, 205

On Human Nature (Wilson), 178

On the Origin of Species (Darwin), viii, 35, 40–41, 45–48, 53, 57, 58, 65, 143, 165, 183, 197, 208

ontogeny, 6

origin of life, 54, 191, 217–19

original sin, 4, 6–13, 32, 60, 215

orthogenesis, 55–57

overpopulation, 70–71

Overton, William, 102–3, 186

Paine, Thomas, 29

Paley, William, 22, 27, 29–31, 33, 37, 155, 160, 163

paradigms, 23–27, 31–32

Patriarchs and Prophets (White), 123

Paul, Apostle, 9

PBS, 82, 166, 176

phylogeny, 6

physics, 4, 5, 13, 185–89

Pinker, Steven, 44, 80, 175, 183

Popper, Karl, 185–89, 193

population growth, 70–71

Price, George McCready, 59, 124–29, 131–34, 137, 142, 151

primates, 13–14, 16, 199, 209

Principles of Geology (Lyell), 33, 49

progress, 54, 55, 57, 68, 78–79

progressive creationism, 60, 132

Provine, William, 182, 183

pseudogenes, 202–4

pseudoscience: Baugh's firmament theory, 147; in the classroom, 95, 109, 115; creationism associated with, 139, 140, 147–48; evolution as, 186, 187, 193; the Hovind Theory, 148

public opinion, 6, 17, 118, 141, 165

public schools debate: affecting academic standards, 94, 130, 140; Arkansas Act 590, 98–103, 108; balancing evolution/creation, 94–98, 103–9; creating false dichotomy, 117–19; defeat of ID curriculum, vii, 17; Dover trial, 111–19, 153; the Epperson case, 94–96; evolutionists as dominating, 141, 144, 149, 165; ID movement strategy, 97, 110–11; NABT statement, 167–69, 176; "No Child Left Behind" program, 169–72; prayer in, 94; scientific creationism banned from, 17, 149; struggles in teaching evolution, 17, 165–69, 183; in the Supreme Court, 106–9; textbook publisher's role in, 82, 93, 94, 97, 129; TRACS accreditation, 141. See also Scopes trial

"Pupkins," 129, 133, 138

purpose of existence, ix, 213–14

Quarks, 217, 220

Race, 75–80

racism, 68, 72, 73, 75, 77, 83

radioactive dating, 5, 122, 127, 139, 142

Ramm, Bernard, 131–38

rape, 80, 82, 173, 174, 183

Reason in the Balance (Johnson), 181

relativity, 170–71, 185, 187, 190

religion (Christian): coexistence with science, vii–viii, 15–16, 28–29, 45–49, 53, 59, 138, 146; concept of hell in, 38–39, 60; conservative Christian intellectuals, 119, 135, 146–47, 149, 153; early 20th-century reformulations of, 59–60; evolution as "dissolving," 9–11, 16–17, 82, 149, 215; fundamental tenets of, 10–11, 40, 59–61; historical power over science, 172–73; as holding on to old proofs, 159, 160, 164; intelligent design as inherent to, 154–56; and liberalism, viii, 2, 6, 46, 59–61; non-friendly debate with science, vii, 28, 128, 136–38, 140–41; opposition to Strauss, 44–45, 59–60; racism in, 77; Roman Catholic tradition, 15, 28, 74, 172, 192; worldview of, 14, 214. See also Bible; evangelical Christians; fundamentalism

reproductive flexibility, 192

retroposons, 203

Reynolds, John Mark, 146, 205

rhea, 33

Rice Institute, 130

Robertson, Pat, 100, 147, 150

Robinson's drugstore (Dayton), 88, 89, 98

Roman Catholic tradition, 15, 28, 74, 172, 192

Ross, Hugh, 50, 205

Rothschild, Eric, 112

Ruse, Michael, 16, 101, 155

Sagan, Carl, 174–77

Santorum, Rick, 153, 169–70, 172

Sarfati, Jonathan, 180

Satan: in Arkansas 590 trial, 101; as deceiving scientists/liberals, 6, 181; evolution as conspiracy of, 20, 38, 44, 83, 140–41, 173, 180

Scalia, Antonin, 109

school system. See public schools debate

science: on age of earth, 5–6, 121, 122, 129; and the Bible, 3, 5, 135–36; concerns about creationism's popularity, 44, 140; gaps in, 159, 164, 216, 220; historic coexistence with religion, vii–viii, 15–16,

28–29, 48, 138, 146; and the meaning of life, ix, 213–14; methods of, 23–24, 157–58, 185–88, 193; naturalism as framework, ix, 159–60, 164; non-friendly debate with religion, 128, 136–38, 140–41; paradigms in, 23–26, 31–32; popularization of, 175–79. See also evolution

scientific creationism. See creationism

scientific literature, vii, 145–46, 166, 176–80, 236n2

Scofield Reference Bible (Scofield), 51–52

Scopes, John, 73–74, 78, 82, 84–86, 88, 89, 91, 93, 96

Scopes trial: Bryan on the witness stand, 50, 90, 91; causes and nuances, 118, 119; facts of the, 88–91; influence on lawsuits following, 91–96, 99, 100, 108, 109; media coverage, 86–87, 91; mythology around the, 85–88

Scopes II, 16, 98–103, 111, 117, 119

secularization/secularism: in America, 86; as brand of scientific community, 128; of creation science, 97–98, 104, 107; Darwin as against, 164; of intelligent design, 111; of science education, 94, 103; as stalled, 183; in Victorian England, 44–46

selfishness, 12

separation of church and state, 93–94, 109

Seventh-day Adventists, viii, 58–59, 61, 122–24, 131, 142

Shockley, William, 79

Simpsons (television series), 143, 175

sin, 12, 174

slavery, 52, 75, 76–77

social Darwinism: definition of, 64, 65; and eugenics, 61, 72–80, 82, 83, 86, 173, 183; as fuel for anti-evolutionists, 65, 82–84, 86, 118, 173; historical framework for, 69–72; negatively coloring biological Darwinism, 61–62, 65–67, 79–81, 173, 183; overview of, 65, 67–69, 81–84; as prescriptive, 64–65, 80; in race relations, 75–79; social units in, 67–68, 80, 81; "survival of the fittest" theme, 64, 65, 68, 71; and unbridled capitalism, 64, 183. See also Nazism

social divide, 91–92, 119, 128

social policy, 73–75

social reform, viii, 55, 61, 83

social units, 67–68, 80, 81
Soviet Union, former, 94, 130
Spencer, Herbert, 65, 69, 71, 183
sperm banks, 75, 79
stars, 5, 25–26, 32, 65–66, 218
stellar evolution, 5, 26, 65–66
sterilization, 61, 73–75, 86, 173
stratified rock, 195–96
Strauss, David Friedrich, 44–47, 59, 60
suffering, 7, 9, 32–35
Sunshine Sketches of a Little Town (Lea-
    cock), 129
"survival of the fittest," 64, 65, 68, 71

Teaching evolution, 17, 165–69, 183, 215
Tennessee anti-evolution laws, 88, 93
Tennyson, Alfred Lord, 66, 161
Teresa, Mother, 213
textbook publishers, 82, 93, 94, 97, 129
"theory" as guess, 103, 108, 112, 128
theory of evolution. See evolution
theory of relativity, 170–71, 185, 187, 190
Thomas More Law Center, 112, 114, 116
Thrasymachus (Greek sophist), 68
thrust faults, 126–27
TRACS (Transnational Association of
    Christian Colleges and Schools), 141

Uber-cell, 54
"Unbiased Comparison of Evolution and
    Creationism" (video), 143
underdetermined theory, 211–12, 220
uniqueness of humankind, 11, 13–15
upland goose, 33, 36–37, 40, 163
U.S. Congress, 169–70
U.S. Constitution, 88, 95, 108
U.S. Supreme Court, 74, 92–93, 104, 106–
    10, 116, 149, 151–52
Ussher, James, 52

Van Till, Howard, 15
vapor-canopy hypothesis, 133–34, 139
Victorian England: admiration of progress,
    54, 55, 57; the church and power, 173;
crisis of faith in, 44–46, 164; skepticism
    of natural selection, 48, 54, 57; social
    struggles in, 69–73, 183
Voltaire (François Marie Arouet), 30, 50

Wallace, Alfred, 36, 48
Ward, Keith, 15, 16
Warfield, Benjamin, ix–x, 138
watchmaker analogy, 30–31
water, 218–19
Weinberg, Steven, 44, 174–76, 178–79, 183
Weldon, John, 19
Wells, Jonathan, 111, 183, 193
Whitcomb, John C., Jr., 2, 5, 6, 59, 124,
    131–37, 139, 142, 145, 151
White, Ellen, viii, 58–59, 122–24, 129,
    137, 142
Wickramasinghe, Chandra, 101–2
William Jennings Bryan College, 91
Williams, David, 102
Wilson, E. O., 174–76, 178, 179, 210
Wise, Kurt, 146, 153, 165, 183, 205
Wonderful Life (Gould), 178
World War I, 61, 173, 183
worldview, 10, 149, 159–60, 166–69,
    180–81
Wright, George Frederick, 60
Wright, Robert, 168

Yom (day), 49–50
young-earth creationism: "brand-name
    creationism," 138–41; as framework
    for creationist texts, 127–28; and the
    Genesis flood, 124, 127, 129, 134–36,
    232n42; and geological column concept,
    124–27; lack of scientific support for, ix,
    128–29, 145–46, 204–5; pre-flood earth
    in, 127, 133–34, 139; public support for,
    121–22, 124, 132; reach and influence
    of, 141–44; Seventh-day Adventists role
    in, 58–59, 122–24

Zimmer, Carl, 82